AF531676

ENCYCLOPAEDIA OF MOLECULAR BIOLOGY

ENCYCLOPAEDIA OF MOLECULAR BIOLOGY

(Vol. 4)

A. Tyagi

ANMOL PUBLICATIONS PVT. LTD.
NEW DELHI - 110 002 (INDIA)

ANMOL PUBLICATIONS PVT. LTD.
H.O.: 4374/4B, Ansari Road, Darya Ganj,
New Delhi-110 002 (India)
Ph.: 23261597, 23278000

B.O.: No. 1015, Ist Main Road, BSK IIIrd Stage
IIIrd Phase, IIIrd Block,
Bangalore - 560 085 (India)
Visit us at: www.anmolpublications.com

Encyclopaedia of Molecular Biology

First Published, 2006

ISBN 81-261-2677-9 (Set)

[Responsibility for the facts stated, opinions expressed, conclusions reached and plagiarism, if any, in these volumes is entirely that of the Author. The Publishers bear no responsibility for them, whatsoever.]

PRINTED IN INDIA

Published by J.L. Kumar for Anmol Publications Pvt. Ltd., New Delhi - 110 002 and Printed at Mehra Offset Press, Delhi.

Contents

Preface

Smallest fundamental unit of an element or a compound is called a molecule. The rapid progress and multi-faceted development of Science, in general; and of Biology, in particular, has ushered in an absolutely new era in the history of mankind, regarding the physio-chemical department of life.

There has been an urge to know more and more from the day one. It is due to this urge that man has been scaling ever new heights day-after-day, be it related to Science or Commerce or Politics and so on. Is this whole endeavour, which has proved to be a real boon for man, or in other words, the greatest gift, the twentieth century gave to humanity ? It really brought about a revolution in the whole human life. The emergence of Molecular Biology, as a branch of Science for study, could also be possible by virtue of this very present of the past.

The discipline is very vast and with the passage of time, big breakthroughs are in the offing. Feeling the pulse of the time, it deemed suitable, rather, necessary to bring out a comprehensive and compact encyclopaedia on this subject in order to cater to the maximum requirements of the lovers of Molecular Biology.

Comprising five volumes, this trivial attempt is expected to prove to be an asset for Science elite and students alike. Now, savour it molecule-by-molecule.

Editor

1

Basic Concepts

A study of basic concepts of genetics and the evidence for chromosome theory of inheritance is useful for students of cell and molecular biology. In this chapter, therefore, basic concepts of genetics (including Mendelism, multiple alleles, polygenes, sex linkage) and chromosome theory of inheritance will be described briefly.

Mendel's Work

Gregor Johann Mendel (1822—1884) described his results on inheritance of seven pairs of traits in pea in the year 1866. This work was rediscovered in 1900 simultaneously by three scientists including Hugo de Vries (a Dutch biologist), Carl Correns (a German botanist) and Eric von Tschermak (an Austrian Botanist). Incidently the term 'Genetics' (used in 1905 by W. Bateson) and 'gene' (coined in 1909 by W. Johannsen) were coined and used only after the rediscovery of Mendel's laws in 1900 as above and also after the formulation of chromosome theory of inheritance in 1902-1903 (by W.S. Sutton and T. Boveri). Mendel used garden pea as plant material for his crossing experiments, since it had the following advantages : (i) well defined characters; (ii) bisexual flowers; (iii) predominantly self fertilization and (iv) easy hybridization. Since garden pea is self-fertilizing, the anthers had to be removed before maturity to facilitate crossing (an operation called emasculation) and pollen brought on the stigma of the female parent from the plant to be used as male parent.

For each of the seven pairs of characters shown in Figure, plants with one alternative trait were used as female and those with the other alternative as male. Reciprocal crosses were also made. The population obtained as a result of crossing plants exhibiting contrasting characters is called the first filial generation or F_1 (filial means progeny). The progeny of F_1 plants obtained due to self-fertilization represents the second filial generation or F_2. Similarly F_3, F_4, etc. could also be obtained. Mendel published the data and conclusions derived from his experiments in 1866 (Table). These results can be explained using a specific example. When tall plants were crossed with dwarf plants, all hybrid plants in F_1 generation were tall. The plants used in the initial cross are referred to as P_1 and P_2 or parents. When these F_1 plants were self-fertilized, both tall and dwarf plants were obtained in the next (F_2) generation. The tall and dwarf plants were obtained in a ratio close to 3 : 1. Similar patterns were obtained for other six pairs of characters also.

The pattern of inheritance observed in all the seven cases can be summarised as follows : (i) For any character the F_1 individuals derived from crosses, between two different varieties having alternative characters, showed only one of the traits and never the other. This feature was expressed as dominance of one trait over the other. The trait which appeared in the F_1 generation was called dominant and the other which did not appear in the F_1 population was called recessive, (ii) It did not matter which parent variety provided the pollen and which provided the eggs; the results were always the same. In other words, the reciprocal crosses gave the same

Actual Data Obtained by Mendel in F_2 Segregating Progenies in Pea

Character studied	Chromosome location (known now)	Dominant	Recessive	Total	Ratio
1. Seed form (*R. r*)	7	5474 (round)	1850 (wrinkled; rugosus)	7324	2.96 : 1
2. Cotyledon colour (I, *i*)	1	6022 (yellow)	2001 (green)	8023	3.01 : 1
3. Seed coat colour (*A, a*) or flower colour (violet *vs* white)	1	705 (grey)	224 (white)	929	3.15 : 1
4. Pod form (*V, v*)	4	882 (inflated)	299 (constricted)	1181	2.95 : 1
5. Pod colour (*Gp. gp*)	5	428 (green)	152 (yellow)	580	2.82 : I
6. Flower position (*Fa. fa*)	4	651 (axial)	207 (terminal)	858	3.14 : 1
7. Stem length (*Le. le*)	4	787 (tall)	277 (dwarf)	1064	2.84 : 1
Total		14, 949	5010	19,959	2.98 : 1

results. The dominant-recessive relationships of the seven pairs of characters are shown in Figure. The determining agent responsible for each trait was called a 'factor'. Since the recessive trait was not seen in the F_1 generation but reappeared in F_2, Mendel could predict the results to be expected in F_3. This will be illustrated with the help of another example.

When plants with yellow seeds were crossed with plants having green seeds, only yellow seeds were obtained in the F_1 generation. This showed dominance of yellow seed colour over green seed colour. When F_1 seeds were sown and seeds were obtained in F_2 generation, these were obtained in a ratio of 3 yellow and 1 green (6022 yellow, 2001 green). Mendel predicted that out of the yellow seeds sown, one-third should give rise to only yellow seeds in the next generation, while the remaining two-third should again give rise to yellow and green seeds in 3 : 1 ratio. The plants raised from green F_2 seeds were expected to produce only green seeds. The results expected by Mendel as outlined in Figure were actually obtained in the F_3 generation for five of the seven pairs of characters studied. Based on his results, Mendel formulated, what are popularly described now as 'Mendel's Laws of Inheritance'. A brief account of these laws follows.

Principle of Segregation

(Law of purity of gametes or Mendel's first law)

Mendel's principle of segregation is inherent in the results of Mendel earlier described in this chapter. This principle suggested that in the F_2 generation, the alternative dominant and recessive traits in a pair segregate. Consequently, the gametes which are formed are always pure for a particular character. A gamete may carry either the dominant or the recessive factor but not both as we find in F_1 individuals. This is why it is called either as 'principle of segregation' or as 'law of purity of gametes'.

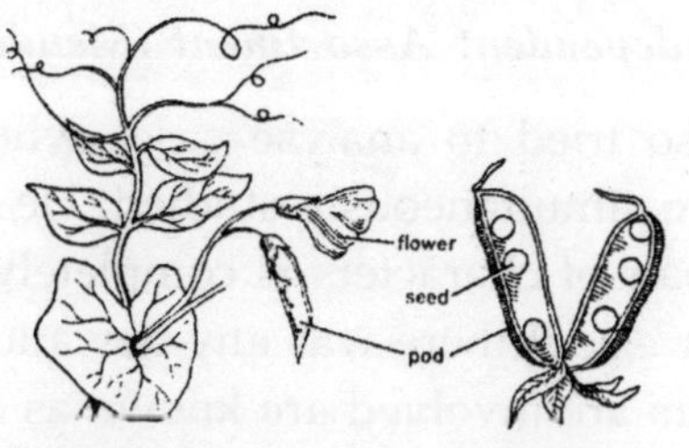

characters	dominant	recessive
seed characters		
shape	round	wrinkled
cotyledon colour	yellow	green
coat colour	gray	white
pod characters		
shape	inflated	constricted
colour	green	yellow
stem characters		
position of pod	axial	terminal
plant height	tall (6–7 ft)	dwarf (3/4–1 ft)

Seven pairs of characters used by Gregor Mendel.

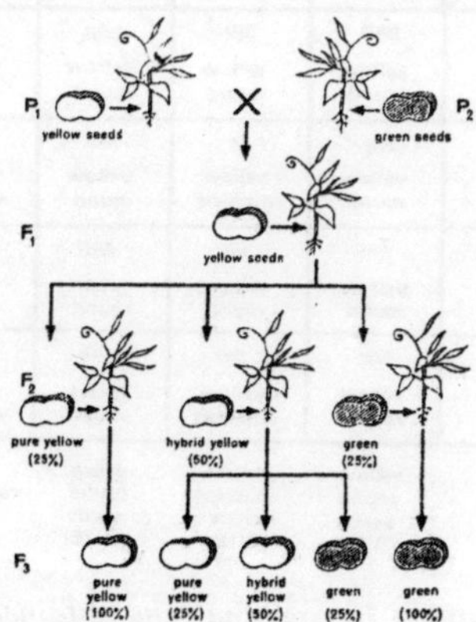

Inheritance of seed colour in pea, showing parents, F_1, F_2 and F_3 generations.

Principle of Independent Assortment (Mendel's second law)

Mendel also tried to analyse cases where two pairs of characters were simultaneously studied. He tried to find out whether one pair of characters is completely independent of another pair or else if there was any deviation. These crosses where two pairs are involved are known as dihybrid crosses. The classical example of a dihybrid cross is the use of two pairs of characters namely the seed shape and seed colour. The plants with yellow and round seeds (pure) were crossed with those having green and wrinkled seeds (pure). The F_1 seeds were yellow and round. When these F_1 seeds were grown into plants, F_2 seeds were obtained which showed all the four

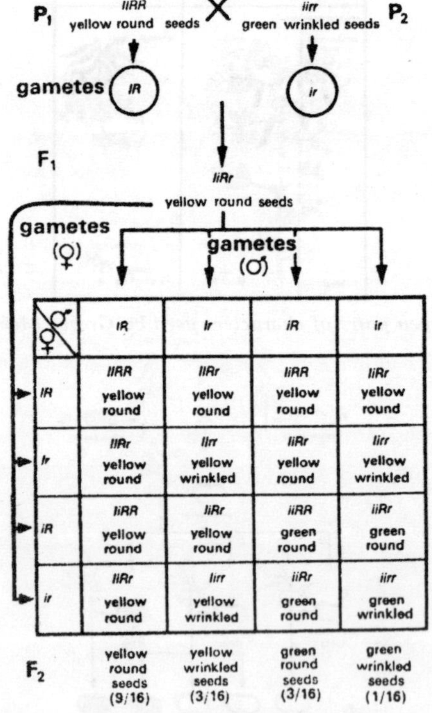

A checkerboard showing 16 combinations obtained in a dihybrid cross involving characters for seed shape and seed colour.

possible combinations i.e. (i) yellow and round seeds, (ii) yellow and wrinkled seeds, (iii) green and round seeds, and (iv) green and wrinkled seeds. These four kinds of seeds were obtained in 9 : 3 : 3 : 1 ratio (Fig.).

The above results would be expected if the two pairs of characters are believed to behave independent of each other. Therefore, it can be concluded that the two characters under consideration are assorting (giving rise to different combinations) in an independent manner. This principle is known as 'principle of independent assortment'. The independent assortment holds good for two or more than two pairs of characters.

Mendel's Results, Chromosome Theory and Linkage (Present Status) : In this chapter, later we will discuss the experiments, which had shown that Mendelian factors (later called 'genes') are located on chromosomes. It will also be shown, that genes located on same chromosome, unless separated by long distances, will not exhibit independent assortment (in other words these will be linked). In view of this we can re-examine the seven pairs of genes used by Mendel and analyse, whether or not they exhibited independent assortment in all possible 21 combinations, when arranged in pairs. This is necessary, because often a parallelism is drawn between seven pairs of characters used by Mendel and the seven pairs of chromosomes found in pea. This is also used as an argument, albeit erroneously, for independent assortment observed by Mendel, by saying that, had he taken more than seven characters, he would have had problems due to linkage. In this connection, we can examine the relationship between different characters with their standard gene symbols and the chromosomes on which each is located (Table).

In his original paper, Mendel gave results of only two experiments showing independent assortment. One of them involved two characters *(R-r, I-i)* and the other involved three characters *(R-r, I-i, A-a)*, such that they represented demonstration of independent assortment among only three of the seven characters in three possible combinations (out of a total of 21 combinations possible among seven characters). For other character combinations, experiments were conducted, but no results are available in Mendel's original paper. However, he claimed that independent assortment was available in all other combinations also.

Extensive linkage studies later conducted by H. Lamprecht (1961), however, demonstrated that seven genes used by Mendel belonged to only four linkage groups. It was also shown that the combinations *i-a, v-fa* and *fa-le* despite each belonging to same linkage group are not linked (due to separation by long distances) and exhibit independent assortment. However, a solitary combination, *v-le* showed linkage (13% recombination) suggesting lack of independent assortment. Therefore, it is obvious that either Mendel did not conduct an experiment for independent assortment between *v* and *le* or else overlooked the lack of independent assortment in this combination.

Molecular Basis of Mendel's Wrinkled Seed Character : In the year 1990, a group of scientists, working at John Innes Institute at Norwich (U.K.), cloned the pea gene *r (rugdsus)*, which determines whether the seed is round or wrinkled. It was shown that an iso-form of 'starch branching enzyme' (SBEI) is present in round *(RR* or *Rr)* seeds, but absent in wrinkled *(rr)* seeds. By careful planned experiments it was shown that there was 100% co-segregation of the absence of SBE I with *r* locus, suggesting that the gene for SBE I is located at *r* locus. It was also shown that in *rr* lines, SBE I gene is interrupted by a small DNA sequence (0.8 kilobases) called 'transposon-like insertion', similar to *Ac/Ds* system in maize. Interruption of SBE I gene by insertion of a foreign sequence, gives an aberrant SBE I enzyme leading to metabolic disturbances in the biosynthesis of starch, lipid and protein.

This results in an increase of free sugars due to failure of starch formation. This probably leads to higher osmotic pressure and hence higher water content and larger cell volume earlier observed in *rr* seeds. The seeds lose large proportion of their volume on maturation leading to shrinkage in volume. Since testa does not shrink with the cotyledons, seeds become wrinkled.

In most cases of the study of Mendelian traits, we seldom know the molecular basis, so that the students do not ask the question : how does a gene control the phenotype? With the availability of recent techniques of molecular biology and recombinant DNA, many traits are now being examined for the molecular basis of their development. This approach is receiving support from the recent emergence of 'reverse genetics' (from DNA to phenotype) as opposed, to 'forward or classical genetics' (from phenotype to DNA).

Mendelian Genetics in Humans

In the above discussion, we utilized the examples of pea as used by Mendel for explaining various aspects of participate inheritance. However, the application of Mendel's laws for a study of problems of heredity in humans will be most interesting, yet it is most difficult for the following reasons (besides others) : (i) controlled crosses can not be made, (ii) duration required for a study of three generations will be long (several decades). In view of the above, human geneticists often scrutinize family histories (pedigrees) in the hope that informative matings might have occurred by chance. This is called pedigree analysis. A member of a family having an exceptional phenotype (e.g. colour blind or deaf or dwarf), which first attracts the attention of a geneticist, is called propositus. The history of the exceptional character in the propositus is traced back in the family and a family tree is prepared using standard symbols (Fig.). Many human diseases

symbol	explanation	symbol	explanation
□	male	■ ● ♂ ♀	affected individuals
○	female	◧ ◐ ♂ ♀	heterozygotes for autosomal recessives
□—○	mating	⊙ ♀	carrier of sex linked recessive
	parents and children (1 boy : 1 girl, in order of birth)		death
	dizygotic twins		abortion or stillbirth (sex unspecified)
			propositus
	monozygotic twins	I 1 2 II 1 2 3	method of identifying persons in a pedigree : here the propositus is child 2 in generation 2 or II 2
◇	sex unspecified		
2 3	number of children of sex indicated	□═○	consanguineous marriage

Symbols used in human pedigree analysis.

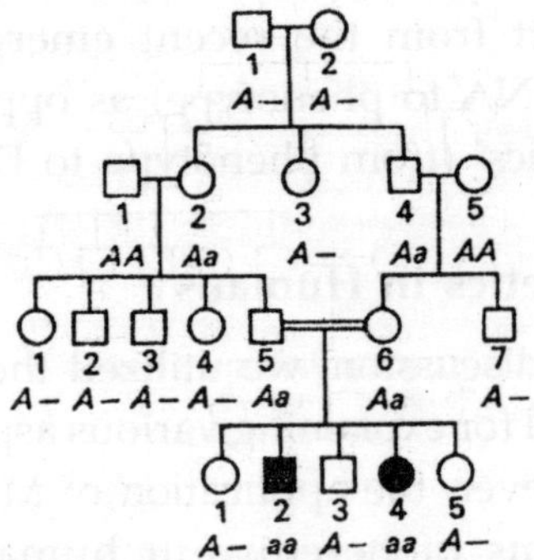

A pedigree involving an exceptional phenotype controlled by recessive allele V (gene symbols are generally not given in pedigrees; they are given here as aid to the readers (redrawn from Suzuki et al., An Introduction to Genetic Analysis, 1986).

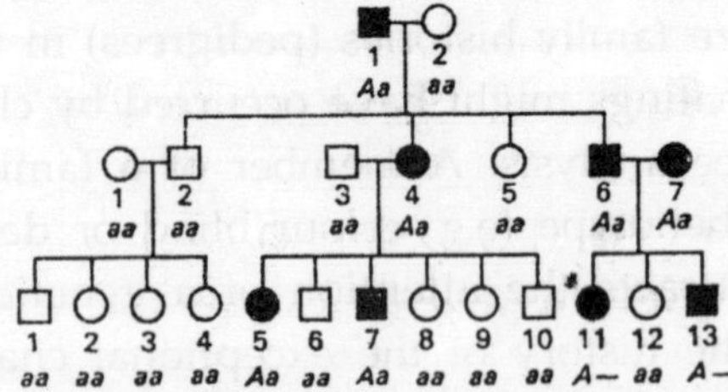

A pedigree involving a rare phenotype controlled by a dominant allele 'A' (redrawn from Suzuki et al.,* An Introduction to Genetic Analysis, *1986).

or defects are governed by simple recessive genes. Common examples of these traits include albino, Tay-Sachs disease, cystic fibrosis, phenylketonuria (PKU), etc. Following generalizations can be made: (i) a recessive character may appear in the progeny of both unaffected parents; (ii) two affected parents can not have an unaffected child; (iii) often these characters appear among children born from consanguineous marriages (*e.g.* first cousin marriage), which enhance the chances of mating among two heterozygotes for the same recessive trait. This also explains why often consanguineous marriages (or sagotra marriages among boy and girl from same gotra—ancestry) are forbidden in several castes in our country. A typical pedigree for a rare recessive condition is shown in Figure.

Rarely, an exceptional character may be governed by a dominant allele (e.g. Huntington's chorea; brachydactyly—very short fingers). This situation will be characterized by the following: (i) the condition occurs in every generation; (ii) unaffected parents can never transmit the condition to the offspring; (iii) two affected parents may have unaffected children; (iv) condition is passed on, an average, to one half of the children of an affected individual. A typical pedigree of a rare dominant condition is shown in Figure.

The pedigree analysis as above, helps in the identification of genes and gives information about the mode of its inheritance. The medical applications of such a genetic analysis, obviously, are far-reaching. This information is particularly useful in day-to-day counselling of prospective parents, who fear genetic disease in their children.

Interaction of Genes

Mendel's experiments suggested that perhaps each character is controlled by a single factor or a single gene.

A Summary of some interactions involving two gene pairs.

Kind of interaction	*Abbreviated genotypes*		*Phenotypic ratio*
1. One gene pair (A), when dominant epistatic on *B*.	*AB* *Ab*	White	12
Example : Fruit colour in summer squash.	*aB*	Yellow	3
	ab	Green	1
A = White, dominant over colour (green) and epistatic on *B*.			
B = Yellow dominant over green.			
Interaction : Dominant *A* (white) masks the effect of yellow or green.			
2. One gene pair *(A)*, when dominant epistatic on second (B)	*AB* *Ab* *ab*	White	13
Example : Feather colour (fowl).	*aB*	Colour	3
A = Colour inhibitor (epistatic on *B)*.			
B = Colour dominant to white.			
Interaction. *A* does not allow *B* to express colour.			
3. Interaction between two dominants to give new phenotype.	*AB*	Disc	9
	Ab *aB*	Sphere	6
Example : Fruit shape in summer squash.	*ab*	Long	1
A, B = Sphere shape dominant over long shape.			
Interaction : *A* and *B*, when present together give disc shape.			

However, a gene does not carry a character, but controls a developmental process. It is through an influence on the development that the gene brings about the desired effect on phenotype. It has also been shown that a gene may sometimes interfere with the development and may cause the death of the individual (such genes are called lethal genes). Similarly, it was also discovered that the genes are not necessarily separate elements producing individually distinct effects. They may interact and produce a phenotype which may be the result of interaction of these genes. Some of these interactions each involving two genes and leading to ratios which are derivations of 9:3:3:1 ratio are presented in Table.

Modifiers, Suppressors and Pleiotropic Genes : Although we normally ascribe the control of phenotype pertaining to a specific character to one particular gene, it is normally influenced in various ways by several genes. The genes, which modify effect of other genes , are sometimes described as modifiers. These modifiers normally change the phenotypic effect of other genes in quantitative manner. Many genes responsible for dilution of body colour may belong to this category. There may always be present modifiers with minor effects on quantitative characters like yield or height, etc. but these genes can not be easily identified.

There are also modifiers, which will not allow mutant allele of another gene to express either fully or partially. These modifiers have been called suppressors and result in wild phenotype, so that one will have to determine whether it is due to reversion of mutant to wild type or due to a suppressor. Some of the examples of suppressor genes are *su-h* (suppressor for hairy wings) and *su-S* (suppressor for star eye shape). Mechanism of action of such suppressor genes.

There are certain other genes, which in addition to their main effect, may also act as modifiers each for another but entirely different gene. Such genes having more than one effect are called pleiotropic genes. Some examples of pleiotropic genes are as follows: (i) Some genes for bristles, eye and wing may influence the number of facets in Bar-eyed individuals, (ii) Gene for white eyes may affect shape of sperm-storage organs in females and also some other structures. It is believed that pleiotropic genes must be more common than earlier suspected, because indirectly every gene may be involved in the expression of more than one traits. Thus most genes, if not all, affect more than one phenotypic traits and are thus pleiotropic in nature.

Meiotic Drive, Segregation Distortion and Selfish Genes : We have discussed above, how Mendelian ratios can be variously modified due to genie interactions, modifiers, suppressors and pleiotropy. There are other cases where certain gametes may be either lethal or may fail to take part in fertilization. This may also lead to modified ratios. One such example is killer gene in wheat which causes death of pollen. Another example is found in *Drosophila,* where a particular class of gametes do not take part in fertilization according to expected meiotic segregation pattern. This phenomenon was called meiotic drive by Sandier and Novitski (1957). For instance certain male flies with genotype *cnbw/cn⁺bw⁺ (en* = cinnabar; *bw* = brown; *cnbw/cnbw* = white due to interaction) when mated with white eyed *cnbw/cnbw* females give rise to a ratio of 25 : 1 (wild = *cnbw/cn⁺bw⁺* : white eyed = *cnbw/cnbw)* instead of expected 1 : 1 ratio.

It has been proved that the above extreme modification of ratio is not due to mortality at zygotic level but is due to the location of a segregation distorter *(SD)* gene carried on *cn⁺bw+* chromosome. It is also established that half the sperms carrying *SD* preferentially fertilize the eggs, so that the genotype *cn⁺bw⁺/ cnbw* predominates. It has also been shown that several loci (at least five) are involved in segregation distortion, located on *SD* chromosome and that there should be a complex mechanism involved (Hartl, 1980).

The *SD* chromosome is found at a frequency of few per cent in most natural populations. The heterozygotes for *SD,* in extreme cases, can produce up to 99% progeny with *SD* due to dysfunction (non-functioning) of gametes carrying the homologue of *SD.* These dysfunctional sperms carrying homologue of *SD* exhibit abnormalities in chromatin condensation (Hartl, 1975, 1980). Segregation distortion has also been noticed in mice where tailless *'t'* allele is recessive lethal. Heterozygotes transmit this allele in frequencies as high as 99% compared to normal allele.

The above two examples i.e. *SD* complex in *Drosophila* and f-locus complex of mice were considered to be the two earliest examples of what were described as selfish genes. These selfish genes were defined as those genes, which propagate themselves, despite being detrimental to the organisms that carry them. The natural selection seems to favour them. Other examples of such selfish genes include B chromosomes, replicative transposons, the psr chromosome of wasps, and mitochondria that cause male sterility in plants. Another recently discovered class of selfish genes includes *Medea* (*M*) in common flour beetle. While, in *SD* and *t* complexes, the allele is transmitted in excess through male (prior to fertilization), in case of *M*, in the progeny of heterozygous mothers (*M* +)," the zygotes which do not carry *M* (homozygotes, + +) die before pupation (post fertilization). Such homozygotes will appear in crosses *M* + × + + or M + × *M* +, but not in crosses when either of the parent is *MM*. The term 'selfish DNA' has also been used sometimes for repeated DNA sequences in a different sense.

Penetrance and Expressivity

We should recognize that a gene can not determine the phenotype in isolation, but its effect on phenotype is controlled both by other genes and by the environment. These factors controlling the expression of a gene are complex and their precise nature can not be established easily. In order to describe this situation, the two useful terms often used in genetics are penetrance and expressivity.

Penetrance is defined as the percentage of individuals with a given genotype, which exhibit the phenotype associated with this genotype. For instance, all individuals with the genotypes *AA* or *Aa*, may not express the dominant phenotype, due to either the presence of modifiers, suppressors, epistatic genes, etc. or due to the modifying effect of the environment. If only 80% of individuals with *AA* or *Aa* express dominant phenotype, it will be described as 80% penetrance of gene (or allele) *'A'*.

Expressivity on the other hand, describes the degree or extent to which a given genotype is expressed phenotypically in a particular individual. This lack of full expression may

	phenotypic expression (each circle represents an individual)
(a) variable penetrance	
(b) variable expressivity	
(c) variable penetrance and expressivity	

Effects of penetrance and expressivity on a hypothetical trait 'pigment intensity'. In each row all individuals have same genotype (redrawn from Suzuki **et al, An Introduction to Genetic Analysis, 1986).**

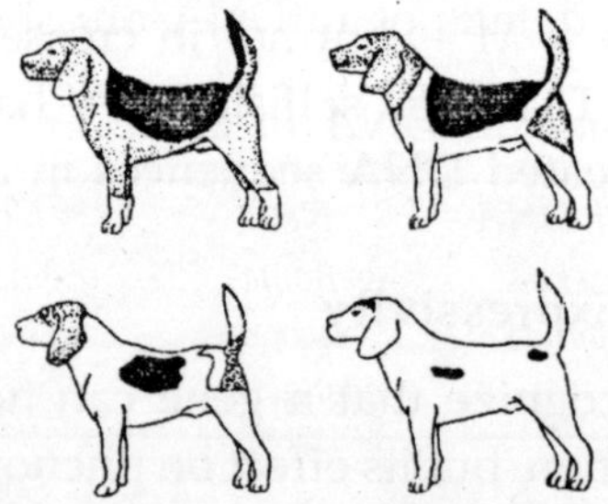

Four grades of piebald spotting in dogs attributed to variable expressivity (redrawn from Little, 1957).

sometimes be due to the rest of genome and the environment. For instance, the red eye colour in *Drosophila* flies may show different shades of colour (within the norm of reaction) suggesting variable expression of wild allele '*W* in *WW* or *Ww* flies. The distinction between penetrance and expressivity is illustrated in Figure, and variable expression of coat colour in dogs is shown in Figure.

Multiple Factors or Polygens

In this Chapter, earlier we discussed cases, where the heterozygous condition of a gene had the same phenotypic effect as the homozygous dominant had. However in some cases the F_1 individual (heterozygous) would show intermediate character with respect to the two parents. These cases earlier

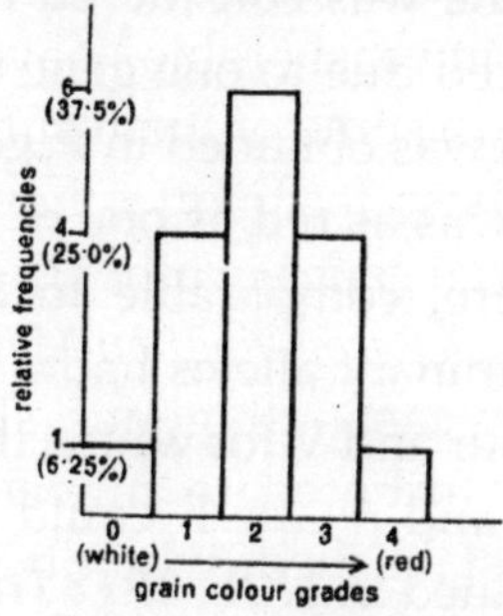

Relative frequencies (theoretical) of different grain colours in wheat in F_2 population derived from two strains differing in two gene pairs.

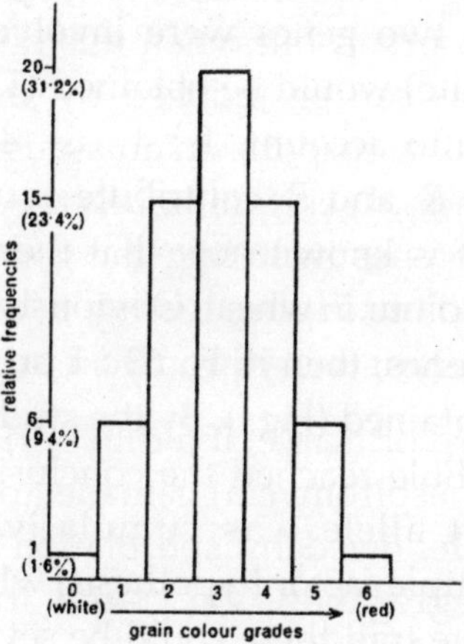

Relative frequencies (theoretical) of different grain colours in wheat in F_2 population derived from two strains differing in three gene pairs.

referred to as cases of blending inheritance are now explained on the basis of incomplete dominance. Such individual characters are each controlled by many genes described as multiple factors or polygens. The inheritance due to multiple factors or polygenes will be illustrated in this section with the help of an example of kernel colour in wheat (studied by H. Nilsson-Ehle).

Kernel colour in wheat is a quantitative character and was studied by H. Nilsson-Ehle for the first time in 1908. It was argued that if one gene was considered or in other words, if the two parents differed due to one gene only, a 3 : 1 ratio for red and white kernels was obtained in F_2 generation. However, out of three red, one was as red as one of the parents and two were lighter and were, comparable to F_1 individuals. This indicated that the dominant alleles had a cumulative effect. If 'R' stands for red colour and V for white, the two parents could be designated as *RR* and *rr*, the F_1 could be designated as *Rr* and F_2 would be obtained in 1*RR* : 2*Rr* : 1*rr* ratio. In these three classes, *RR* should be red, *Rr* should be intermediate in colour and rr should be white.

In case, where two genes were involved, a 15 : 1 ratio (15 coloured : 1 white) would be obtained (Table). If different shades are taken into account, 1 : 4 : 6 : 4 : 1 ratio will be obtained, provided R_1 and R_2 contribute equally to the colour (Fig.). However, it is known now that there are three genes involved in kernel colour in wheat. Obviously if the two parents differ in all three genes, then in F_2, 63 : 1 or 1 : 6 : 15 : 20 : 15 : 6 : 1 ratio will be obtained (Fig.). By the study of kernel colour in wheat, Nilsson-Ehle reached the conclusion that the effect of each dominant allele was cumulative. Therefore, he forwarded his multiple factor hypothesis, which states that for a, given quantitative trait there could be several genes, which were independent in their segregation, but had cumulative effect on phenotype.

Genotypic and Phenotypic Ratios in F_2 Obtained Due to a Cross $R_1R_1R_2R_2 \times r_1r_1r_2r_2$.

Genotype	Genotypic ratio	Phenotype	Phenotypic ratio	
$R_1R_1R_2R_2$	1	Red (like P_1)	1	Coloured (15/16)
$R_1R_1R_2r_2$	2	Darker than F_1 and lighter than P_1	4	
$R_1r_1R_2R_2$	2			
$R_1r_1R_2r_2$	4	Like F_1 (Intermediate between P_1 and P_2)	6	
$R_1R_1r_2r_2$	1			
$r_1r_1R_2R_2$	1			
$R_1r_1r_2r_2$	2	Lighter than F_1	4	
$r_1r_1R_2r_2$	2			
$r_1r_1r_2r_2$	1	Colourless	1	Colourless (1/16)

Multiple Alleles and Isoalleles

Mendel's concept of inheritance envisaged alternative forms of same gene or factor. This would mean that each gene had two alternative forms or allelomorphs, one being dominant or partially dominant and the other recessive; one being wild form and the other mutant. If the mutant form developed from wild form due to mutation, there is no reason, why we should not expect that the wild form can mutate in more than one ways.

The mutant form can also mutate once again to give rise to another mutant form. Therefore, a gene can have more than two allelomorphs. These aflelomorphs make a series of multiple alleles. Relationship between different alleles may or may not be a simple dominant-recessive relationship, but they were usually, characterized by their location on the same locus, and therefore, by lack of crossing over between them.

However, discovery of pseudoalleles has rendered this definition out of date and it is known now that multiple alleles, can occupy different sites within a gene and that crossing over between them can take place.

It is also possible that several alleles may exhibit themselves within the same phenotypic range, whether wild or mutant. Such alleles expressing themselves within the same phenotypic range are called isoalleles. If the phenotype is wild, these are normal or wild isoalleles and if the phenotype is mutant, they will be called mutant isoalleles. Such examples are many and one of them is noticed in eye colour in *Drosophila*, where several alleles exhibit red eye colour (W^{+s}, W^{+c}, W^{+g}). The phenomenon of multiple allelism will be illustrated with the help of a popular example, of ABO blood groups in humans.

On the basis of presence of absence of certain antigens, ABO blood groups have been established in man by K. Landsteiner. It was found that there can be two antigens A or B and as a result four blood groups. These blood groups are called ABO blood groups or Landsteiner blood groups after the name of the discoverer.

With the antigens A and B, there are certain naturally occurring antibodies in the serum of the blood. General principle of antibody and antigen relationships is, that antibodies in a particular individual will be found only for those antigens which are absent in blood of this individual. The status of antigens and antibodies in different blood groups is shown in Table.

Different ABO Blood Groups and the Antigens and Antibodies Associated with them

Blood groups	*Antigen antigen*	*Antibody for*
A	A	B
B	B	A
AB	A and B	None
O	None	A and B

Antibodies in blood group A will be able to agglutinize red blood corpuscles of the blood group B and *vice-versa.* AB blood group will not agglutinize any other group, since no antibodies are present.

Similarly, O blood group should be able to agglutinize all other three groups except its own. On this relationship is based the test whether a particular blood group can be donated for another specific blood group. Table shows the compatible donor and recipient groups.

Compatible Donor and Recipient ABO Blood Groups

Blood group of donor	*Blood group of recipient*
A	A and AB
B	B and AB
AB	AB (universal recipient)
% O (universal donor)	O, A, B, AB

Table suggests that, if donor's blood has antibodies against that of recipient, it does not matter, since they get diluted. However, there should be no antibodies in recipient's blood against donor's blood, since this will cause agglutinization of transfused blood.

It would also be obvious, that AB is universal recipient and no test of donor's blood in such case is necessary. Similarly, O group is universal donor and no test for the recipient's blood is necessary, if blood of 0 group is available for transfusion.

In ABO blood groups discussed above, antigens (also called agglutinogens) are present in the red blood cells and antibodies (also called agglutinins) in the serum.

Consequently, transfusions can be made after examining the compatibilities of blood groups, so that the transfused blood may not get agglutinated. As shown, the blood group

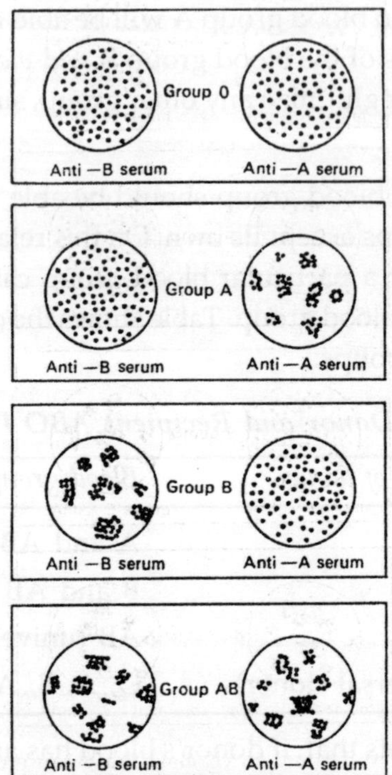

Effects of anti-A (containing antibody A) and anti-B (containing antibody B) sera on corpuscles of four blood groups (O, A, B and AB). A clumped pattern within a circle shows agglutinization of red blood corpuscles.

A contains anti-B serum (antibodies for B antigen) and blood group B contains anti-A serum. The effects of anti-A serum and anti-B serum on agglutinization of different blood, groups are shown in Figure.

Different ABO Blood Groups and their Genotypes

Blood group	*Possible genotypes*
O	I^OI^O or *ii* or ++
A	I^AI^A or *AA*; I^AI^O or I^A *i* or *A*/+
B	I^BI^B or *BB*; I^BI^O or I^B *i* or *B*/+
AB	I^AI^B or *A*/*B*

The genetics of the above blood groups indicates that three alleles are present: (i) I^0 or i or + (ii) I^A or A and (iii) I^B or B. I^A and I^B are mutant alleles and are dominant over the wild allele I^O or i or +. This is based on the concept that wild type is one which is more frequent. O and AB blood groups are more frequent and if O is assumed as wild, it yields a convenient system. The genotypes of different blood groups are shown in Table.

Sex Linkage

Earlier in this chapter, while discussing the laws of inheritance, it was realized that reciprocal crosses would give same results. In other words, it did not make any difference, whether a particular character was present in male parent or in female parent. For instance, in pea when tall plants are crossed with dwarf plants, always tall plants are obtained in F, generation, irrespective of whether tall plants are used as male or as female parent. This would always be the situation in bisexual plants or in hermaphrodite animals. However, in dioecious individuals, there can be two kinds of characters, namely: (i) characters which do not show any differences in reciprocal crosses (♀A×:♂ B; ♀B×♂ A) and (ii) characters which show a difference in reciprocal crosses. The genes for the former type of characters are located on autosomes and those for the latter type are located either on specialized chromosomes known as sex chromosomes, or are located in the cytoplasm. Characters of this latter type, the genes for which are located on sex chromosomes, are described as sex linked and will be illustrated using the example of colour blindness in humans. The characters, for which genes are located in the cytoplasm are classified under maternal effects or cytoplasmic inheritance.

A particular trait in human beings renders them unable to differentiate between red colour and green colour. The gene for this red green colour blindness is located on X-chromosome.

Colour blindness is recessive to normal vision so that if colour blind man marries a girl who is normal (homozygous) for this character, sons will be normal, but daughters will be heterozygous (normal phenotype), which means that these daughters would be earners of this trait.

If such a carrier girl marries a colour blind man, 50% of female progeny and 50% of male progeny would be colour blind.

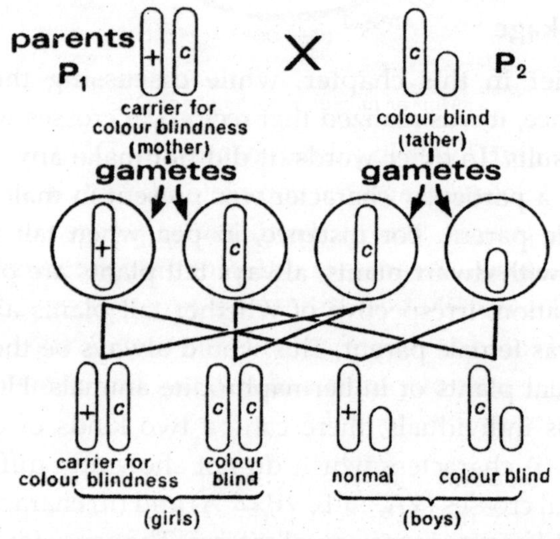

Inheritance of sex linked colour blindness in human beings.

Non-disjunction of sex chromosomes in Drosophila

Primary Non-disjunction : In 1916, C.B. Bridges reported that in rare cases (one in 2,000 or 3,000), results contrary to expectations could be obtained in inheritance experiments on sex linked characters. For instance, when *red eyed* (XY=w^{+}–) males are crossed with *white eyed* (XY= ww) females, one expects white eyed males and *red eyed* females in F_1 generation.

However, Bridges found that rarely in such a cross red eyed males (X-=w^{+}) and white eyed females (XXY=ww-) were observed. Bridges explained this instance on the basis of non-disjunction of X-chromosomes in female individual.

The phenomenon of non-disjunction means absence of separation of two homologous X-chromosomes during anaphase I of meiosis. Instead, both X-chromosomes go together to same pole. Other pole obviously will receive no X-chromosome. Results obtained due to such an event are shown in Figure. Such a non-disjunction with normal female parent would be called primary non-disjunction in order to distinguish it from secondary non-disjunction.

Secondary Non-disjunction : The XXY *(ww-)* females obtained due to primary non-disjunction were again used in further experiments and secondary non-disjunction was shown.

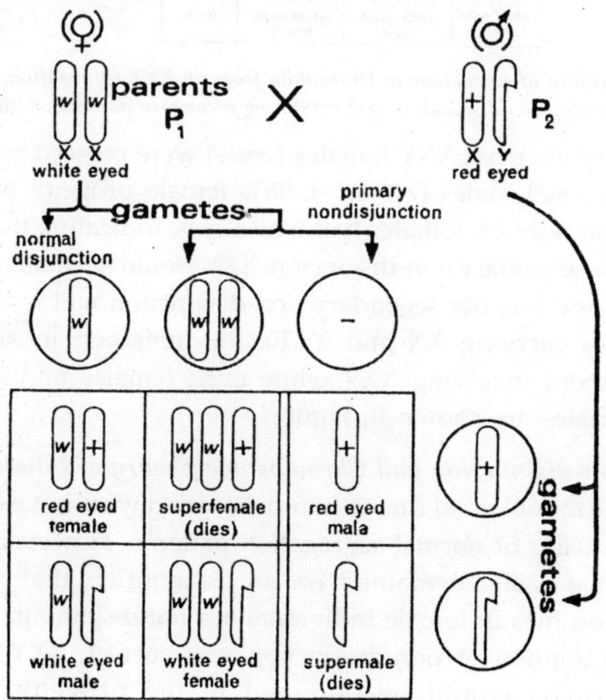

Inheritance of eye colour in Drosophila due to primary non-disjunction of sex chromosomes.

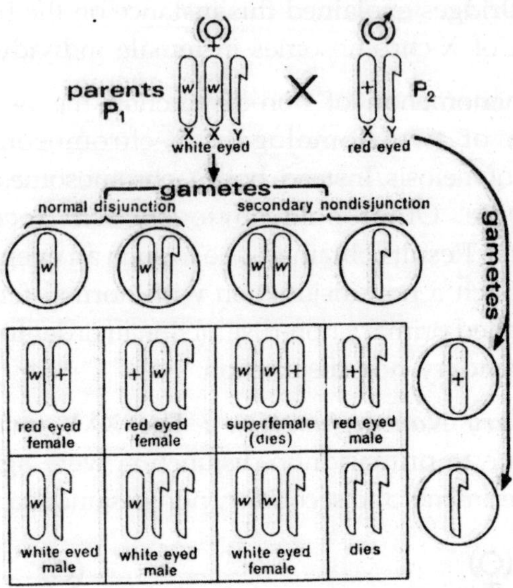

Inheritance of eye colour in Drosophila from an XXX fly resulting from primary non-disjunction and exhibiting secondary non-disjunction.

When white eyed XXY females (ww–) were crossed with red eyed normal males (XY=w^{+}--), 96% female progeny had red eyes and only 4% females had white eyes, indicating that their was 4% secondary non-disjuncion. XXY would normally disjoin into X and XY, but secondary non-disjunction will result into gametes carrying XX and Y. Results obtained in such an experiment involving XXY white eyed females and XY red eyed males are shown in Figure.

Non-disjunction and Chromosome Theory of Inheritance : The results obtained due to non-disjunction were not expected on the basis of normal segregation patterns. However, these could be easily explained on an assumption that two X-chromosomes in female individual are not disjoining. If such an assumption of non-disjunction was correct, XXY female individuals would be obtained in the progeny. When exceptional females (white eyed) were cytologically studied, these were found to carry two X-chromosomes and one Y-chromosome (XXY). Such cytological results definitely suggested and, therefore, were put forward as an evidence to show that the genes or Mendelian factors were carried on chromosomes. Therefore, non-disjunction of sex chromosomes (primary and secondary) is considered as a strong evidence in favour of the chromosome theory of inheritance.

Chromosome Theory of Inheritance : The chromosome theory of inheritance mentioned above, was actually formulated much before the above evidence in its support was presented by C.B. Bridges. Soon after the rediscovery of Mendel's laws in 1900, W.S. Sutton (an American graduate student) and T. Boveri (a German biologist), in 1903, independently showed a parallelism between behaviour of chromosomes and Mendelian characters. The similarities included the following and are shown in Figure. (i) The chromosomes occur in pairs like the alleles of a gene, (ii) The homologous chromosomes separate during meiosis like the pair of similar or dissimilar alleles of a gene separate at the time of gamete formation, (iii) Different chromosomes orient and separate independently during meiosis, showing parallelism with Mendelian factors.

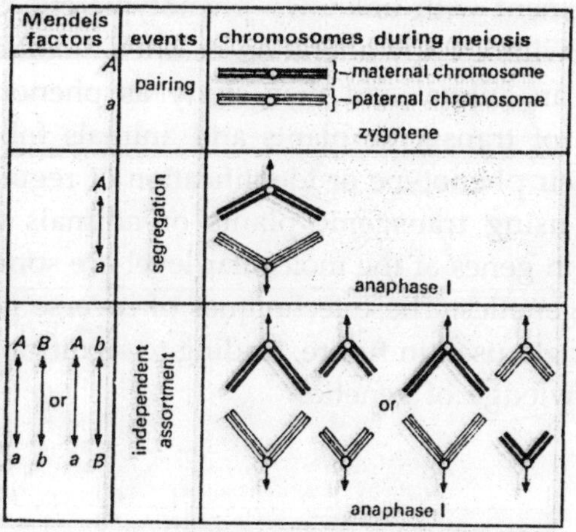

Parallelism between Mendel's hypothetical particles (genes) and chromosomes during meiosis.

It was therefore, concluded that genes or the Mendelian factors were located on chromosomes.

Forward Genetics vs Reverse Genetics : Mendelian genetics described in this chapter is sometimes described as forward genetics, where we start the study on the basis of phenotype, leading ultimately to the study of DNA sequences comprising the gene for this phenotype. In contrast to this, during the last two decades the term 'reverse genetics' has been frequently used for a discipline involving mapping and isolation of genes, whose protein products are not known. Even though the mapping and isolation of a gene for any trait (including those for human diseases) with unknown gene product is a spectacular achievement, but the approach can not be termed reverse genetics for the simple reason that in this case also, we start the genetic study on the basis of phenotype and finally isolate the gene, albeit using sophisticated techniques of recombinant DNA.

Therefore, Paul Berg (Nobel Laureate) in a later report (1991) suggested that the usage of the term reverse genetics be restricted to those studies, where we start the study with a DNA segment with unknown phenotypic effect, introduce this DNA (without any alteration or after modification) into a plant or an animal and then study its phenotypic effect. Production of transgenic plants and animals followed by a study of their phenotype or identification of regulatory DNA sequences using transgenic plants or animals or targeted alterations in genes at the molecular level are some examples of reverse genetics. These techniques of reverse genetics will be increasingly used in future, leading to significant advances in our knowledge of genetics.

2

Genetic Code

We know that the genetic information in the nucleic acids is expressed through synthesis of proteins. We also know that, while proteins have 20 different kinds of essential amino acids, nucleic acids (particularly in living calls) have only four different kinds of bases. (Recently, it has been shown that in artificially synthesized DNA segments, the number of bases can extend to more than four *i.e.* 6 or 8).

It may therefore, be realized that while 20 amino acids are the alphabets of the language of proteins, four bases are the alphabets of the language of nucleic acids. Since it is through mRNA (messenger RNA) that the genetic information is passed on to proteins, the problem of genetic code was to prepare a dictionary for translating the language of RNA into the language of proteins.

Since one is four alphabet language and the other has 20 alphabets, as shown in Figure, a singlet code would give only four codons, a doublet code would have 16 codons and triplet code would give us 64 (4 × 4 × 4) triplets. The 64 triplets would be enough to code for 20 amino acids.

In a 64-codons dictionary, either there should be more than one triplets coding for the same amino acid or else, of the 64 triplets, 44 triplets should mean nothing (nonsense). It has now been proved by experimental evidence that the code is triplet and that all 64 codons carry some meaning, although three of these 64 triplets cannot be translated in terms of amino acids, since they mean stop signals. Whenever anyone of these three triplets is encountered on mRNA, it orders that the end of the polypeptide chain has reached and that the chain should now be released, before further synthesis may take place.

singlet code
A
G
C
T

doublet code			
AA	AG	AC	AT
GA	GG	GC	GT
CA	CG	CC	CT
TA	TG	TC	TT

triplet code			
AAA	GAA	CAA	TAA
AAG	GAG	CAG	TAG
AAC	GAC	CAC	TAC
AAT	GAT	CAT	TAT
AGA	GGA	CGA	TGA
AGG	GGG	CGG	TGG
AGC	GGC	CGC	TGC
AGT	GGT	CGT	TGT
ACA	GCA	CCA	TCA
ACG	GCG	CCG	TCG
ACC	GCC	CCC	TCC
ACT	GCT	CCT	TCT
ATA	GTA	CTA	TTA
ATG	GTG	CTG	TTG
ATC	GTC	CTC	TTC
ATT	GTT	CTT.	TTT

A singlet code, a doublet code and a triplet code.

Properties of Genetic Code

Following properties of the genetic code were proved by definite experimental evidence : (i) the code is triplet, (ii) the code is degenerate, (iii) the code is non-overlapping, (iv) the code is commaless, (v) the code is non-ambiguous and (vi) the code is universal. Although it may not be necessary to present the experimental evidences which proved the validity of these properties, it may be useful to explain the meaning of these six properties of the genetic code listed above.

The Code is Triplet : As earlier outlined, singlet and doublet codes are not enough to code for 20 amino acids; it was pointed out that triplet code is the minimum required. But it could be a quadruplet code or of a higher order. As pointed out above, in a triplet code of 64 codons, there is an excess of 44 codons and, therefore, more than one codons are present for the same amino acid. This excess will be still greater if more than three-letter words are used. In the quadruplet code there will be 4 × 4 × 4 × 4 = 256 possible words.

The Code is Degenerate : In a triplet code, as pointed out earlier also, for a particular amino acid more than one words (synonyms) can be used. This phenomenon is described by saying that the code is degenerate. A non-degenerate code would be one where there is one to one relationship between amino acids and the codons, so that 44 codons out of 64, will be useless or nonsense codons. It has been definitely shown that there are no nonsense codons. The codons which were earlier called nonsense codons are also now known to mean stop signals.

The Code is Non-overlapping : Non-overlapping code means that a base in a mRNA is not used for two different codons. In Figure it is shown that an overlapping code can mean coding for four amino acids from six bases. In actual practice six bases code for not more than two amino acids. It is shown that the same base can be used for different codons,

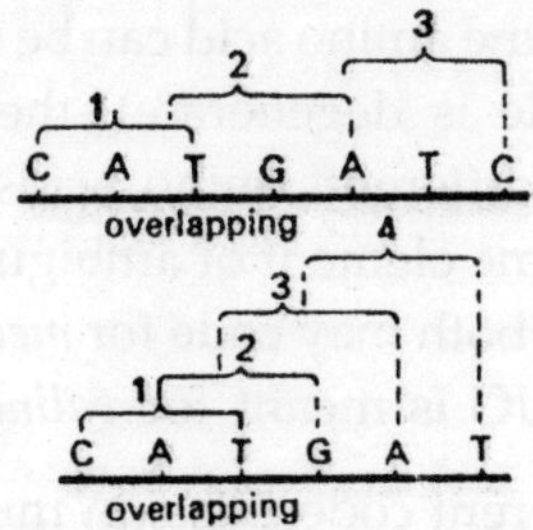

Overlapping of codons due to one letter or two letters.

but only at different occasions in time and/or space, so that the same base can not be used for two different codons during synthesis of the same protein. This is exactly, what we mean by non-overlapping code.

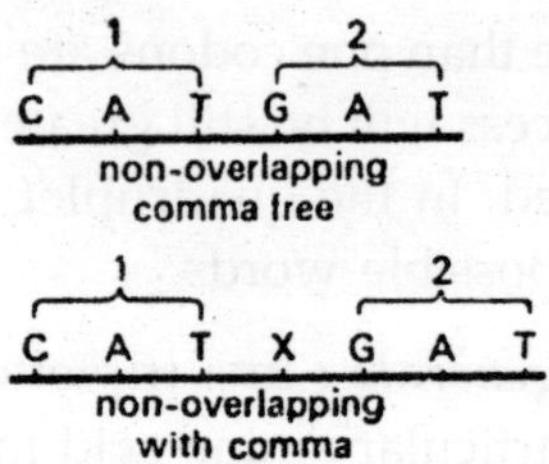

Genetic code, without comma and with comma.

The Code is Commaless : A commaless code means that no punctuations are needed between any two words. In other words, we can say that after one amino acid is coded, the second amino acid will be automatically coded by the next three letters and that no letters are wasted for telling that one amino acid has been coded and that now second should be coded (Fig.).

The Code is Non-ambiguous : Non-ambiguous code means that there is no ambiguity about a particular codon. A particular codon will always code for the same amino acid, wherever it is found. In an ambiguous code, the same codon could have different meanings, or in other words, the same codon could code two or more than two different amino acids. Such is not the case. While the same amino acid can be coded by more than one codons (the code is degenerate), the same codon shall never code for two different amino acids (non ambiguous). However, there is some element of ambiguity when AUG and GUG are considered; both may code for *methionine* as initiating codons, although GUG is meant for *valine* (Fig.).

Moreover, a different code exists in mitochondria of some eukaryotes, so that in cytoplasm and mitochondria same codon may code for different amino acids. This may not mean ambiguity, since in mitochondria there is a separate genetic code which differs from universal code in some essential respects.

The Code is Universal **:** Although the code has been worked out by using *in vitro* systems prepared usually from micro-organisms, there is no doubt now that in all kinds of living organisms, micro- or macro-, plants or animals, the same genetic code is used.

Codon Assignments

Although theoretical considerations in 1950's had suggested that the genetic code should be triplet in nature, it was not possible to say which codon of the possible 64 codons should code for which of the 20, essential amino acids. The first clue to this problem came when M.W. Nirenberg (Nobel Prize winner with H.G. Khorana and R.W. Holley in 1968) used *in vitro* system for the synthesis of a polypeptide using an artificially synthesized mRNA molecule. The use of artificial mRNA molecule has a definite advantage, since one could know something about its structure.

Assignment of codons with unknown sequence Polyuridylic acid or poly-U and other homopolymers. In the first place, M.W. Nirenberg and J.H. Matthaei synthesized RNA using only uracil, so that all along the length of mRNA, there was no other base and the only possible triplet was UUU. When such a poly-U (RNA) was used in cell free synthesis of a polypeptide (using cell extracts from *E. coli* and supplying all the required components of protein synthesizing machinery), only polyphenylalanine was synthesized, meaning that the only amino acid coded was phenylalanine.

It was, therefore, immediately concluded that the triplet UUU coded for the amino acid *phenylalanine.* This news that the genetic code had been cracked splashed in the newspapers all over the world in 1961. Subsequently, poly A gave *polylysine* and poly C gave *polyproline*. Therefore, AAA was assigned to *lysine* and CCC to *proline*. Similar experiments with poly G were not successful, because it attains secondary structure and thus cannot attach to ribosomes. The following three codons thus could be assigned, using homopolymers of RNA, without any difficulty; UUU = *phenylalanine;* AAA = *lysine;* CCC = *proline.*

Copolymers : Having successfully used the homopolymers, Nirenberg and his co-workers used RNA synthesized by using two or more bases. For instance, if only A and C are used, poly AC will consist of eight possible codons namely AAA, AAC, ACA, CAA, CCA, CAC, ACC and CCC. The proportion of these eight codons in the synthetic RNA can be calculated if the known quantities of A and C are used for the synthesis of poly AC. For instance, if A : C = 5 : 1, (5/6 is A and 1/6 is C), the calculated relative proportions of eight codons on random basis would be as given in Table.

The Relative Proportions of Different Codons in mKNA Formed Due to Bases A : C taken in 5 : 1 Ratio

Base composition	*Codon and probability*	*Ratio using* min. (as 1)	*Ratio using* max. (as 100)
(1) 3A	1. AAA 5/6 x 5/6 x 5/6 =125/216	125	100
(2) 2A1C	2. AAC 5/6 x 5/6 x 1/6 = 25/216	25	20
	3. ACA 5/6 x 1/6 x 5/6 = 25/216	25	20
	4. CAA 1/6 x 5/6 x 5/6 = 25/216	25	20
(3) 1A2C	5. CCA 1/6 x 1/6 x 5/6 = 5/216	5	4
	6. CAC 1/6 x 5/6 x 1/6 = 5/216	5	4
	7. ACC 5/6 x 1/6 x 1/6 = 5/216	5	4
(4) 3C	8. CCC 1/6 x 1/6 x 1/6 = 1/216	1	0.80

The calculated relative proportions of codons were compared with the proportions in which different amino acids were present in the polypeptides synthesized using poly AC. For instance, if an amino acid is 1/5th of lysine (coded by AAA), we can say that it should be coded by one of the three possible 2A1C codons (AAC, ACA or CAA).

Similar reasoning would allow assignments of 1A2C as well as 3C. However, using this technique, it was not possible to assign the three codons of the category 2A1C (*i.e.* AAC, ACA, CAA) to three amino acids, since these will be present in equal quantities. Therefore, the codons were initially assigned only with respect to base composition, ignoring the sequences of the bases in codons, as done in the above example. The assignments are given in Table.

Codon Assignments Derived Due to Use of A : C (5 : 1) in the Synthesis of mRNA.

Amino acids	*Codon composition*
Lysine	3A
Asparagine, glutamine & threonine	2A1C
Histidine, proline & threonine	1A2C
Proline	3C

Assignment of Codons with Known Base Sequence

A large number of synthetic copolymers of unknown sequences was used for preparing a dictionary where codon compositions were assigned to amino acids.

The next problem and a serious one was to sort out the three codons in one base composition *e.g.* 2A1C or 1A2C. In other words, we can say that the next problem was to work out the sequences of bases in codons, whose composition was already known. This was done mainly by the following methods.

Binding Technique of Nirenberg and Leder : M.W. Nirenberg and P. Leder in 1964 found that if a synthetic trinucleotide for a known sequence (with known bases at 5' end and 3' end) is used with ribosome and a particular aminoacyl-tRNA (tRNA having its own specific amino acid attached), these will form a complex, provided the used codon codes for the amino acid attached to the given aminoacyl tRNA.

$$\text{Codon}_1 + \text{Ribosome} + \text{AA}_1 - \text{tRNA}_1 \rightarrow \text{Ribosome} - \text{Codon}_1 - \text{AA}_1 - \text{tRNA}_1$$

In a process such as above, if given AA_1 is used with a given $codon_1$ and the formation of the complex is detected, this would prove that the given codon codes for the given amino acid.

It was also observed that while the free AA-tRNA passes through nitrocellulose membrane, the ribosome-codon-AA-tRNA complex adsorbs on such a membrane. If in a particular mixture only one of the amino acids is made radioactive, then the presence or absence of the radioactivity on the nitrocellulose membrane will show whether there is a relationship between the codon and the amino acid which was made radioactive.

For instance, 20 samples of a mixture of all 20 amino acids may be taken and in each sample one amino acid is made radioactive in such a manner that each and every amino acid is made radioactive in one sample or the other, and no two samples have same radioactive amino acid.

A particular sample would be then known by its radioactive amino acid. Now tRNAs and ribosomes are mixed with each sample and same codon is used for complex formation in all 20 cases. When the mixture is poured on the nitrocellulose membrane, radioactivity on membrane will be observed only when the radioactive amino acid is taking part in the formation of the complex.

Since in each sample the radioactive amino acid is known it would be possible to detect the amino acid coded by a given codon by the presence of radioactivity on the membrane.

Such a treatment was given by Nirenberg and his co-workers to all the 64 synthetic codons, and their respective amino acids were identified. The binding of AA-tRNA was not equally efficient in all cases. Therefore, the sequences of bases in only about 45 codons could be worked out by this method.

Copolymers of Repetitive Sequences by H.G. Khorana : As outlined above, sufficient information regarding the sequences of bases in codons was available through the painstaking work of Nirenberg and co-workers.

However, H.G. Khorana, almost at the same time, devised an ingenious technique for the same purpose. Using synthetic DNA, Khorana and his co-workers could prepare polyribo-nucleotides (RNA) with known repeating sequences. A repeating sequence means that, if CU are two bases, these will be repeatedly present throughout the length as follows :

CUCUCUCUCUCUCU

In a similar manner, if ACU are three bases they will be present repeatedly as follows :

ACUACUACUACU

Such copolymers will direct the incorporation of amino acids in a manner which can be theoretically predicted. For instance, in (CU) – (CUC/UCU/CUC/UCU), only two codons

Assignment of Codons, having kn Sequences, with the Help of Copolymers have Repetitive Sequences of Two Bases

Copolymers used	*Codons in copolymer*	*Amino acids incorporated*	*Codons assigneo*
$(CU)_n$	CUC/UCU/CUC	leucine/serine	CUC/UCU
$(UG)_n$	UGU/GUG/UGU	cysteine/valine	UGU/GUG
$(AC)_n$	ACA/CAC/ACA	threonine/histidine	ACA/CAC

Assignment of Codons, having known Sequences, with the Help of Copolymers having Repetitive Sequences of Three Bases = $(ACG)_n$.

Codons	*Homopolypeptide*	*Codon assignment*
ACG/ACG/ACG/ACG/ACG = Poly (ACG)	$(\text{Threonine})_n$	ACG = threonine
A/CGA/CGA/CGA/CGA/CGA = Poly (CGA)	$(\text{Arginine})_n$	CGA = arginine
AC/GAC/GAC/GAC/GAC = Poly (GAC)	$(\text{Aspartic acid})_n$	GAC = asartic acid

possible and these are CUC and UCU. Moreover, these codons are present in alternating sequence.

The result would be that the polypeptide formed would have only two amino acids in alternating sequences. These two amino acids can be assigned to the two codons (Table).

We may similarly consider a repeating sequence of three bases *e.g.* $(ACG)_n$. Depending upon where the reading is started, three kinds of homopolypeptides are expected (Table).

Actual codon assignment *i.e.* to find out which of the three codons codes for which amino acid would depend upon the previous information available regarding the composition of bases in different codons coding for different amino acids.

On the basis of the above techniques, a complete genetic code dictionary could be prepared which is given in Figure. It can be easily noticed that there are two codons namely AUG and GUG, which are designated as initiation codons and there are three codons UAA, UAG and UGA which are designated as termination codons.

Chain Initiation and Chain Termination Codons : In most if not all polypeptides synthesized *in vivo,* the initiating amino acid is methionine. The corresponding codon on mRNA is usually AUG, but rarely in could be GUG also. When GUG is used

First letter	Second letter: U		C		A		G		Third letter
U	UUU	Phe (F)	UCU	Ser (S)	UAU	Tyr (Y)	UGU	Cys (C)	U
	UUC		UCC		UAC		UGC		C
	UUA	Leu (L)	UCA		UAA	stop	UGA	stop SeCys	A
	UUG		UCG		UAG	stop	UGG	Trp (W)	G
C	CUU	Leu (L)	CCU	Pro (P)	CAU	His (H)	CGU	Arg (R)	U
	CUC		CCC		CAC		CGC		C
	CUA		CCA		CAA	Gln (Q)	CGA		A
	CUG		CCG		CAG		CGG		G
A	AUU	Ile (I)	ACU	Thr (T)	AAU	Asn (N)	AGU	Ser (S)	U
	AUC		ACC		AAC		AGC		C
	AUA		ACA		AAA	Lys (K)	AGA	Arg (R)	A
	AUG*	Met (M)	ACG		AAG		AGG		G
G	GUU	Val (V)	GCU	Ala (A)	GAU	Asp (D)	GGU	Gly (G)	U
	GUC		GCC		GAC		GGC		C
	GUA		GCA		GAA	Glu (E)	GGA		A
	GUG*		GCG		GAG		GGG		G

Genetic code dictionary; three letter and one letter abbreviations are used for amino acids

as the initiating codon, it does not code for valine (as shown in Fig.). However, initiating methionine molecule should be found in the formylated state. This makes a distinction between the initiating methionine and the methionine at internal position. The methionine when required at internal position should not be formylated. Also while formyl methionine is carried by $tRNA_f^{met}$, there is a separate species of tRNA for internal methionine and it is designated as $tRNA_m^{met}$. It should also be realized that although the genetic code *in vivo* is unambiguous (no codon coding for more than one amino acid) there is an element of ambiguity with GUG, since if present at initiating position, it would code for methionine, but would otherwise always code for valine.

As shown in Figure, UAA (also known as ochre), UAG (also known as amber) and UGA (also known as opal) are three codons which whenever present in mRNA, would bring about termination of polypeptide chain which would then be released from the ribosome. Till the function of these codons (UAA, UAG, UGA) was discovered, they were called nonsense codons, since these codons did not code for any of the 20 essential amino acids. Now that we know, that they have a definite function of termination to perform, it would not be correct to call them nonsense codons.

Synonym Codons and Degeneracy

As earlier pointed out and as can be seen from Figure, there are more than one codons for each of several amino acids (see Table). The frequency of different amino acids in proteins also agrees with the number of codons representing each amino acid (Fig.), thus suggesting that the synonym codons have been retained in the genetic code for a definite purpose and is not a wasteful phenomenon. It may also be seen that codons representing same amino acid are clustered in the genetic code dictionary (Fig.), and that the synonym codons usually differ in the third base only, a phenomenon sometimes called third base degeneracy (Table).

Mutations and the Genetic Code

Most of the work earlier discussed in this chapter deals with the study of genetic code in cell free systems. Therefore, it could be questioned whether or not this information would apply to the living systems. With the study of certain mutants,

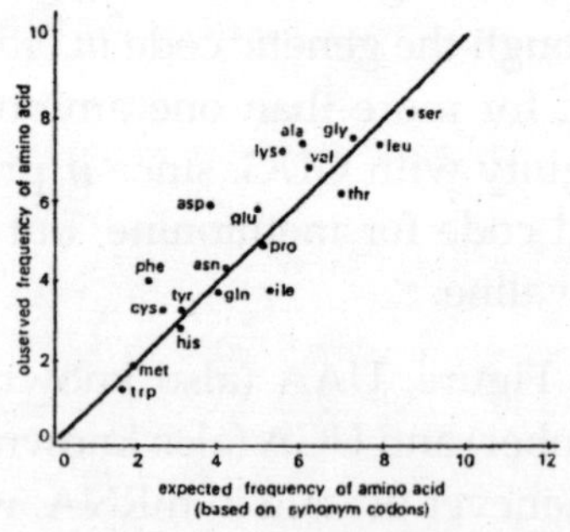

Relationships of frequencies of codons for different Amino Acids in the Genetic code (this gives expected frequencies of amino acids in proteins) and the observed frequencies of amino acids in proteins of living systems (drawn from Lewin's "Genes VI").

Amino Acids Coded by One or more Codons in Genetic Code Dictionary

Number of codons	*Amino acids*
One	*Met,* (initiating amino acid), *Trp.*
Two	*Phe, Tyr, His, GluN, AspN, Lys, Asp, Glu, Cys.*
Three	*Ileu.*
Four	*Val, Pro, Thr, Ala, Gly.*
Six	*Ser, Arg, Leu.*

it was possible to show that the genetic code deciphered using cell free systems applies to the living systems also. There are two kinds of mutations which played a very

Frequencies of Codons having Third base Degeneracy

Third base having same meaning	*Number of codons*	
U, C, A, G	32	
U, C, A	3	(Ileu)
U, C	14	purine is distinguished from pyrimidine
A, G	12	" " " "
G only	2	(AUG = Met; UGG = Trp)
A only	1	(UGA = termination)

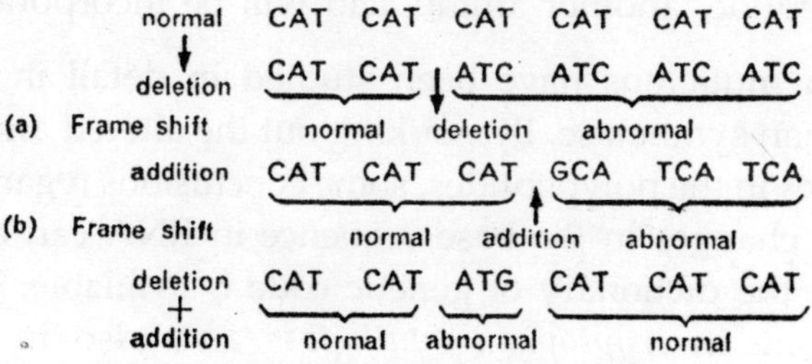

Frame-shift mutations due to deletion and addition.

significant role in the study of the genetic code in living systems. These are (i) frameshift mutations and (ii) base substitutions.

Frameshift Mutations : The genetic message, once initiated at a fixed point, is read in a definite frame in a series of three letter words. The framework would be disturbed as soon as there is a deletion or addition of one or more bases. There is evidence that mutations induced by acridine dye and related compounds depend on addition or deletion of single bases. When such frameshift mutations were intercrossed, in certain combinations they gave wild type.

It was concluded that one of them was deletion and the other an addition, so that the disturbed order of the frame due to one mutation will be restored by the other (Fig.). In such a restoration, a few amino acids coded by mRNA segment lying between the sites of deletion and addition would differ from normal protein. These differences could be explained on the basis of the Genetic Code and, therefore, are used as an evidence in favour of the properties described for Genetic Code.

Base Substitutions or Amino Acid Replacements : If in a mRNA at a particular point, one base pair is replaced by another without any deletion or addition, the meaning of one codon containing this altered base will change. Consequently, in place of a particular amino acid at a particular position in a polypeptide, another amino acid will be incorporated.

Such mutations have been studied in detail in enzyme tryptophan synthetase. By working out the altered amino acid sequences in the polypeptides, some conclusions regarding the possible changes in the base-sequence in RNA can be easily made, if the dictionary of genetic code is available. This has been done in *tryptophan synthetase* and also in case of haemoglobin using sickle cell anemia. All such mutations could be easily explained on the basis of the Genetic Code dictionary given in Figure. The mutations thus provide an evidence to suggest that the Genetic Code deciphered in cell free system holds good in living cells also.

Wobble Hypothesis

It is sometimes argued that the third base of a codon is not very important and that specificity of a codon is particularly determined by the first two bases. It has been shown, for instance that the same tRNA can recognize more than one codons differing only at the third position. This pairing is not very stable and is allowed due to wobbling in base pairing at this third position. Dr. F.H.C. Crick, in 1965 proposed a hypothesis called *Wobble hypothesis* to explain this phenomenon. He discovered that if U is present at first position of anticodon, it can pair with either A or G at the third position of codon. Similar is the case with G and I (1 = inosine is a modified base in tRNA) found in anticodon. The pairing relationships between first base of anticodon and third base of codon are given in Table).

Wobble Pairing.

Anticodon (first base)	Codon (third base)
U	A, G
C	G
A	U
G	U, C
I (inosine, resembles G)	U, C, A

This kind of wobbling allows economy of the number of tRNA molecules, since several codons meant for same amino acid are recognized by same tRNA. For instance, anticodon IGC can recognize codons GCU, GCC and GCA (coding for alanine); IGA can recognize UCU, UCC and UCA (coding for serine) and IAC can recognize GUU, GUC and GUA (coding for valine). In a primitive code discovered in mitochondria, the extent of wobbling is further increased, so that only 22 instead of 55 tRNAs are needed.

Differences between Universal Genetic Code and Mitochondrial Genetic Code

	Universal genetic code	*Mitochondrial code (mammals, yeast)*
1.	55 anticodons (RNA)	22 anticodons (RNA)
2.	3 termination codons (UAA, UAG, UGA)	4 termination codons (UAA, UAG, AGA, AGG)
	(a) UGA = termination	UGA codes for tryptophan
	(b) AGA and AGG code for arginine	AGA and AGG are termination codons
3.	CUN (N = any nucleotide) codes for leucine	CUN codes for threonine in yeast mitochondria
4.	UAG anticodon is aminoacylated by leucine	UAG anticodon in yeast mitochondria accepts threonine

New Genetic Codes in Mitochondria and Ciliate Protozoa: Earlier we described that the genetic code is universal. It was also believed that the genetic code does not undergo any kind of evolution and, therefore, should be static. However, during 1980s, it was shown that variations in genetic code are found in mitochondria, particularly studied in yeast and mammals. In these cases (yeast mitochondria), UGA codes for tryptophan, although in the nuclear genes, UGA is a termination codon (Fig.).

It has also been shown that although a new genetic code may exist in mitochondria, but mitochondria in all organisms may not have the same genetic code. For instance, UGA in mitochondria, does not always code for tryptophan. In the gene coding for maize cytochrome oxidase subunit II, UGA codon is not present and tryptophan is coded by codon CGG (coding for arginine in the nuclear genes; see Fig.), in addition to the codon UGG. There are also indications that AUA (coding for isoleucine in nuclear genes (see Fig.) codes for methionine in mitochondria of mammals, *Xenopus*, yeast, and *Drosophila*, but not in those of *Neurospora* and *Aspergillus*.

The CUN triplets (CUA, CUG, CUU, CUC) in yeast mitochondria, but not in some fungi, code for threonine instead of leucine. In mitochondria of mammals, but not in fungi, there are four initiating codons (AUG. AUA, AUU, AUC) instead of two (AUG, GUG) in the universal code, and they are recognized by N-formylmethionyl-tRNA as in bacteria, but unlike eukaryotes.

second base		second base		third base
C	UCU	U	UUU	U
	UCC		UUC	C
	UCA		UUA	A
	UCG		UUG	G
A	UAU	G	UGU	U
	UAC		UGC	C
	UAA		UGA	A
	UAG		UGG	G

An archetypal code, showing 16 (sixteen) anticodons, with first base (5′) always U (uracil) which can wobble and pair with any of the four bases at 3′ end of mRNA triplet codon

Another unique feature of mitochondrial code of mammals and *Xenopus,* is the use of AGA and/or AGG as stop codons rather than as arginine codons. In mitochondria of *Drosophila* AGA appears to code for serine. It is thus obvious that there is considerable diversity in the genetic code of mitochondria (Table).

The need for a different genetic code in mitochondria perhaps arose due to small genome size in mitochondria. This necessitated an evolutionary retrogression towards a more primitive code. In 1966, T.H. Jukes presented the concept of primitive code called archetypal code, in which one anticodon (in tRNA) will pair with a family of four codons due to wobbling at the first base of anticodon and third base of codon. This four way pairing will provide for only 15 amino acids instead of 20, since there are 16 anticodons (Fig.) and one of them will have to be excluded to provide for chain termination. However, in mammalian mitochondria (Mmt) this archetypal code is modified since provision needs to be made for 20 amino acids.

anti codon	amino acid	anti codon	amino acid	anti codon	amino acid	anti codon	amino acid
GAA	phe	UGA	ser	GUA	tyr	GCA	cys
UAA	leu			—}stop		UCA	tryp
UAG	leu	UGG	pro	GUG	his	UCG	arg
				UUG	gluN		
GAU	ileu	UGU	thr	GUU	aspN	GCU	ser
CAU	met			UUU	lys	—}stop	
UAC	val	UGC	ala	GUC	asp	UCC	gly
				UUC	glu		

Anticodons in the mammalian mitochondrial genetic code; anticodons in each box are meant for four mRNA triplet codons shown in Fig. For instance, anticodon GAA pairs with codons UUU and UUC, anticodon UAA with codons UUA and UUG, anticodon UAG with CUU, CUC, CUA and CUG and so on.

Sequences of complete mitochondrial DNA are now known in human, bovine, mouse and some other organisms. Each mitochondrial DNA has been shown to have genes for 22 tRNAs as against 55 tRNAs in universal genetic code discovered in *E. coli* and applicable in most cases. These 22 tRNAs are sufficient to read codons for 20 amino acids and can pair with 60 codons, remaining four codons being used for stop signals in Mmt. This reduction in the number of tRNAs is achieved by four way pairing and two way pairing. Four way pairing is used eight times (32 codons, 8 amino acids, 8 tRNAs); G pairs with U and C six times and U pairs with A and G another six times (24 codons; 12 amino acids; 12 tRNAs). This provides for 20 amino acids with 56 codons (using 20 tRNAs).

Of the remaining eight codons, there are two each for leucine and serine using two additional tRNAs, since both of them have six codons each and there are four stop codons (UAA, UAG, AGA, AGG). These 22 anticodons and their corresponding 20 amino acids are given in Figure. The differences between universal code and mitochondrial code are given in Table, and the diversity in mitochondrial code in several organisms is shown in Table.

It will be interesting to compare Figure with Figure. The major differences include the following : (i) one termination codon (UGA) and two arginine codons (AGA, AGG) from

Differences in Meaning of Several Codons in Universal Code and Mitochondrial Genetic Code in a Few Organisms

Organism	*Codon*	*Probable meaning in mitochondrion*	*Usual meaning*
All cases	UGA	Tryptophan	Termination
Yeast	CUA	Threonine	Leucine
Fruitfly	AGA	Serine	Arginine
Mammal	AGA }	Termination	Arginine
Mammal	AGG }	Termination	Arginine
Mammal	AUA	Methionine	Isoleucine

universal code have different meanings in the mitochondrial genetic code ; UGA in Mmt (mammalian mitochondria) coding for tryptophan and AGA and AGG in Mmt coding for stop signals; (ii) number of tRNAs is 22 in Mmt code, while it is 55 in *E. coli*. In Table it is also apparent that the mitochondrial code in mammals and yeast may differ in some respects.

In 1986, a different genetic code (like in mitochondria) has been shown to be present in ciliate protozoa *(Mycoplasma capricolum).* In this genetic code, codons UAA and UAG specify glutamine instead of stop signals. In future, more such cases may come to light among primitive organisms.

Suppression Mutations, Base Substitutions and Suppressor tRNAs : When the effect of one mutation is suppressed by another mutation, the latter is called a suppressor mutation. For instance, *AABB* may be wild, and *aaBB* a mutant; then if *aabb,* is also a wild type, the mutation in *'B'* will be called suppression mutation, because the effect of *'a'* is suppressed by *'b'*. Mechanism of this suppression effect was not known for some time, but later its mechanism became largely known. These effects can be conveniently classified into (i) intragenic suppression, and (ii) intergenic suppression.

Intragenic Suppression : Suppression of a mutation at one site within a gene can be achieved by mutation at another site within the same gene. For instance, a mutant A46 that produces inactive tryptophan synthetase enzyme, due to substitution of

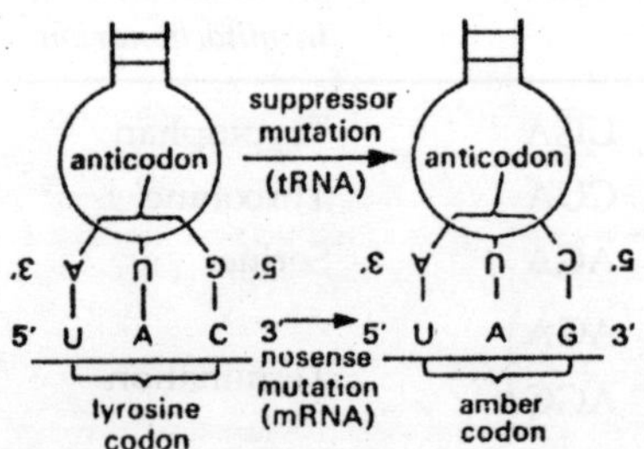

Suppressor mutation involving a change in one base of anticodon of tyrosine tRNA leading to reading of termination amber codon UAG, which resulted from nonsense mutation in tyrosine codon UAC thus leading to suppression.

glutamic acid for glycine, at position 210 may be corrected or suppressed due to mutation A446 involving substitution of cysteine in place of tyrosine at position 174. These two mutants A46 and A446 independently give rise to inactive enzyme, but if present together give rise to active enzyme, thus suppressing the mutant effect of each other. In many other combinations, double mutants do not show any suppression effect. The suppression effect really depends on whether or not the second alteration can restore the 3-D configuration of the enzyme. There are a number of known examples of such intragenic or intracistronic suppression or complementation. The reader is advised to consult an advanced textbook for further details in this connection.

Intergenic Suppression : Intergenic suppression, which means suppression of mutation in one gene due to mutation in another gene, can be achieved mainly due to alteration in transfer RNA (tRNA). The suppression may also be effective by alteration brought about in ribosome conformation. There are at least four types of such suppression mutations, which can be discussed.

Nonsense Suppression and Mutant tRNA : Sometimes due to a change in anticodon of a specific tRNA, it may develop the capacity to read the mutant terminating codon (amber or UAG, ochre or UAA and opal or UGA mutants). Since these nonsense mutations cause premature termination of polypeptide chain, they can be suppressed if a mutant tRNA

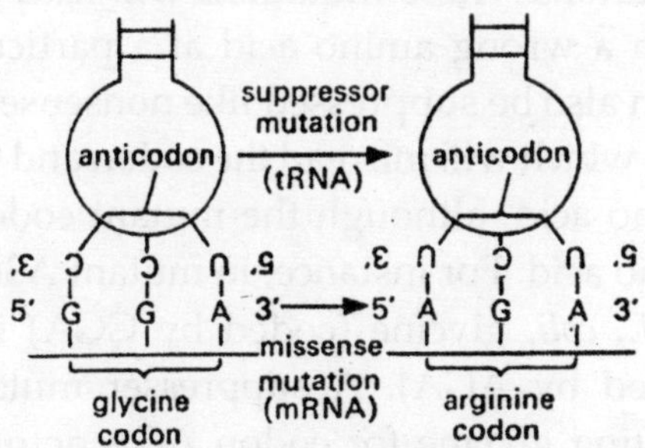

Suppressor mutation involving a change in one base of anticodon of glycine tRNA leading to misreading of arginine codon AGA (as glycine) which resulted due to missense mutation in glycine codon GGA thus leading to suppression.

can read the terminating codon and substitute an amino acid at the location of termination codon. Mutant tRNAs are known which are capable of recognizing termination codons and insert amino acids like *tyrosine, serine* or *glutamine* to restore the function of protein (Fig.). However these mutant tRNAs may also read the normal stop signals of proteins thus causing elongation of protein beyond its normal termination thus causing disturbances. To prevent this disturbance, double stop signals are often available in mRNA, so that if mutant tRNA reads one, the other may still bring about termination. Moreover, two genes may code for same tRNA, so that if one (minor gene) mutates, and gives rise to suppressor tRNA the other (major gene) still functions and gives rise to normal tRNA.

Transfer RNA (tRNA) mediated UAG suppression was first observed, when in coat protein of phage R17, amber codon 5′UAG3' could be read by a tyrosine suppresser tRNA (Fig.), in which the anticodon had mutated from 5′GUA3' to 5′CUA3'. There are also other examples of this type of suppression.

Missense Suppression and Mutant tRNA : We have already discussed mutations resulting due to base substitutions leading to replacement of one amino acid by another. These are called missense mutations. These mutations will lead to synthesis of a protein with a wrong amino acid at a particular site. Such a mutation can also be suppressed like nonsense mutations, by mutant tRNA, which will misread the codon and thus substitute a correct amino acid, although the mutant codon codes for a different amino acid. For instance, in mutant A36 in *tryptophan synthetase* of *E. coli,* glycine (coded by GGA) is replaced by arginine (coded by AGA). A suppresser mutant may have tRNA, depositing glycine for codon AGA actually meant for arginine (Fig.).

Frameshift Mutations and Mutant tRNA : Earlier in this chapter these mutations were discussed, where the reading frame of mRNA shifts to the right due to deletion or to the left due to addition of a base pair in DNA. It has been shown that a frameshift mutation due to a deletion can be suppressed by mutant tRNA having an anticodon of two bases rather than three (doublet anticodon rather than triplet). Similarly, addition frameshift mutations can be suppressed by mutant tRNAs each having a quadruplet anticodon instead of a triplet. Mutant tRNAs with doublet and quadruplet anticodons have actually been detected in biological systems.

Suppression and Ribosomal Mutations : Mutations in some ribosomal proteins may cause a change in ribosome structure, so that it may now misread the mRNA codons. One such mutation, called *ram* (ribosomal ambiguity) is capable of reading all the three termination codons, thus causing suppression of nonsense mutations. Some frameshift mutations may also be suppressed by these *ram* mutants.

Mutations may also occur in ribosomes rendering them sensitive to an antibiotic like streptomycin. These sensitive ribosomes may then allow misreading of mRNA codons, although in resistant ribosomes this misreading is much less frequent. For instance, when Poly-U is used as a template with sensitive ribosomes, most frequent error is replacement of phenylalanine (coded by UUU) by isoleucine (coded by AUU). It has been shown that streptomycin sensitive strains carry mutations in 30S-ribosomal proteins, causing suppression through misreading.

'Second Genetic Code' and 'Second Half of the Genetic Code' : A distinction, therefore, needs to be made of these two expressions from *'The Genetic Code'* described in this chapter. As already mentioned earlier in this chapter, the genetic code is meant for translation of the language of nucleic acid into that of protein, using triplet codons. However, the 'Second Genetic Code' refers to the mechanism involved in the interaction of specific tRNAs with corresponding amino acyl synthetase enzymes (how does an enzyme recognize its tRNA or *vice-versa,* and the 'Second Half of the Genetic Code' refers to the protein folding problem or to the rules, which govern the formation of a three dimensional structure from the primary structure (amino acid sequence) of a protein. Both these problems remain only partly resolved and continue to receive attention of molecular biologists.

X-ray crystallographic structures of several 'tRNA-synthetase complexes' representing intermediates in the formation of 'amino-acyl tRNAs' have been studied. The most important of these is a complex between 'glutamyl-tRNA synthetase' with $tRNA^{Gln}$ (tRNA for glutamine) and ATP. It was shown that there are multiple points of contact all along the inner side of the L-shaped structure of tRNA, which helps in recognition of tRNA by synthetase enzyme and in the formation of linkage between tRNA and amino acid.

Recoding of the Genetic Code

In recent years, examples have been discovered, where in a minority of cases, there is a set of instructions (called 'reading signals') contained in the mRNA itself, that specify an alteration in how the genetic code is to be applied. These alterations are described as 'recoding' and include (i) alteration in the linear read out mechanism, involving + 1 or – 1 shift (or a hop of many nucleotides) in the frame of reading, at a particular point, when mRNA is being read (meaning of codon words remains unaltered); (ii) alteration in the meaning of the code words.

Alteration in the Linear Read-out Frame : There are several examples of this recoding mechanism : (i) In *E. coli,* mRNA of RF2 (protein meant for termination of translation = release factor) programmes 30% of ribosomes to change to the + 1 reading frame after codon number 25 (codon 26 is UGA; see Fig.). The frameshift event is encouraged by an upstream sequence (termed 'stimulator') that pairs with 16S RNA in the ribosome. There are two components of recoding signals, a site of action (codons 25 and 26) and a stimulator. (ii) In a class of retroviruses, many mRNAs exhibit a ribosomal frame shift to the — 1 reading frame by tandem slippage of tRNA. The two components of 'recoding signal' are, a site of action consisting

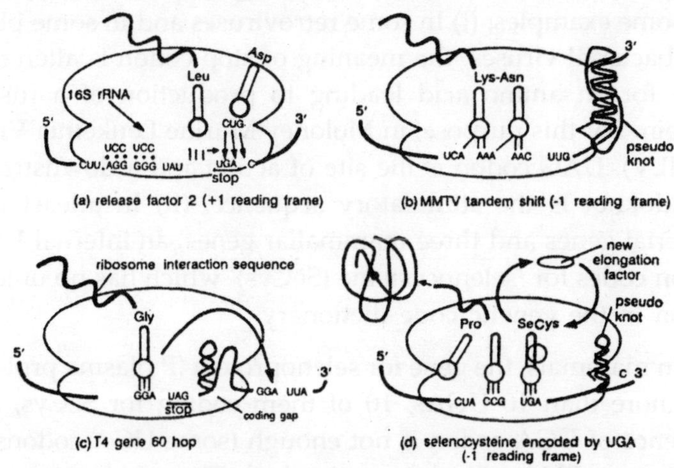

Different alternatives leading to recoding of the genetic code.

of a heptanucleotide in the mRNA and a stimulatory sequence in the form of stem loop or pseudoknot structure (present downstream; see Fig.). A similar heptanucleotide for frameshift is found in *dnaX* gene of *E. coli*, where 50% of gene product in foreshortened due to frameshift. (iii) In higher animals, a protein called 'antizyme' renders the enzyme 'ornithine decarboxylase (ODC)' unstable. Decoding of antizyme mRNA requires a + 1 frameshift, which is regulated by the concentration of polyamines (product of ODC). (iv) In the mRNA for T4 *gene* 60, ribosomes hop a stretch of 50 nucleotides found between a pair of glycine codons.

At the first glycine codon, 100% ribosomes are released; these ribosomes hop and read the second glycine codon downstream and continue translation of the rest of mRNA (Fig.). The stimulatory signals include the mRNA structure and the amino acid sequence of nascent peptide chain.

Alteration in the Meaning of Code Words : In almost all organisms, there seem to be available 'recoding signals' in mRNAs leading to altered meanings of codons. These are exceptions of the universal character of genetic code. Following are some examples: (i) In some retroviruses and in some plant and bacterial viruses, the meaning of stop codon is altered to code for an amino acid leading to production of a fusion protein. For this purpose, in Moloney Murine Leukemia Virus (MMLV), UAG codon is the site of action and a downstream pseudoknot is the stimulatory sequence. (ii) In atleast two bacterial genes and three mammaliar genes, an internal UGA codon codes for 'selenocysteine (SeCys)' which has no unique codon in the genetic code dictionary.

In mammals, the gene for selenoprotein 'P plasma protein' has more than 10 UGAs, 10 of them coding for SeCys, but presence of UGA codon is not enough (some UGA codons in the same mRNA code for termination). The requirements for UGA to code for SeCys include a specific *minor tRNA*, a specific *'elongation factor* (EF)' and a particular *'downstream sequence'* (sometimes this downstream sequence is 200 nucleotides away in the 3' non-coding region; see Fig.).

3

Genetic Engineering

All living systems are made up of independent units of life, and so called "cells", and each individual cell consists of cytosol and several membrane-bound compartments, each filled with concentrated aqueous solution of a variety of molecules. The study of these molecules gave birth to the discipline of molecular biology. Molecular biology thus is a discipline to study how living organisms and the cells constituting them are made from molecules. It also tells us how these molecules perform at the cellular level to make an individual organism as simple as acellular viruses and as complex as the multicellular organisms like human beings. Molecular biology has also facilitated the study of how life in this universe might have actually originated in the remote past.

The progress in the discipline of molecular biology during 1950s and 1960s was so spectacular that no area of pure or applied biology remained unaffected by this new approach. During 1970s, the developments in the field of molecular biology also led to the development of a series of exciting techniques, which allowed manipulation of genetic material, the deoxyribonucleic acid (DNA). This new and exciting discipline, which had its roots in the relatively older discipline of molecular biology, has been variously referred to as "genetic engineering", "gene cloning", "*in vitro* genetic manipulation" or "recombinant DNA technology". In practice, this new discipline of "genetic engineering" involves manipulation of genetic material and introduction of this manipulated genetic material into a cell. There, it replicates faithfully and is passed on to progeny cells, thus becoming a permanent component of the hereditary material of the genetically altered cells.

In some cases, these genetically altered cells are used to obtain genetically modified organisms (GMOs). Major discoveries/events, which led to the development and growth of the disciplines of molecular biology and genetic engineering are listed in Table. Many of these discoveries also led to the award of Nobel Prizes in the fields of Medicine/Physiology, Chemistry and Physics (Table). The definitions, a brief history and scope of these two disciplines and the relationship between them will be briefly described in this chapter.

Some Major Discoveries/events, Which Led to the Development and Growth of the Sisciplines of Molecular Biology and Genetic Engineering

Year	*Event*
1777	Macquer classified all substances, which coagulated on heating as "albuminous".
1807	J.J. Berzelius classified matter into 'organic substances' (derived from living organism) and 'inorganic substances'.
1837	G.J. Mulder worked out a basic formula for substance called "albuminous" by Macquer and named it 'protein' (meaning "of first importance").

Contd...

Year	*Event*
1868	Friedrich Miescher isolated DNA for the first time.
1901-1907	Emit Fischer successfully achieved condensation between amino acids; he called small synthetic chains as 'peptides' and the 'CO-NH' group as 'peptide link'.
1920	Svedberg developed an 'ultracentrifuge'
1952	Purcell and Bloch developed NMR.
1930	CV Raman discovered 'Raman effect', meaning that light scattered by molecules in solution had altered wave length.
1944	Avery et al., provided evidence that DNA (not protein), carries the genetic information during bacterial transformation.
1953	Watson and Crick proposed the double-helix model for DNA structure on the basis of X-ray diffraction results obtained by Franklin and Wilkins.
1957	Kornberg discovered DNA polymerase, which is now used for labeling DNA probes.
1961	Marmur and Doty discovered DNA renaturation, the phenomenon utilized for nucleic acid hybridization.
1962	W. Arber provided evidence for existence of DNA restriction nucleases, which were later (1973) purified and used in DNA sequence characterization by D. Nathans and H. Smith.
1960-66	Nirenberg, Ochoa and Khorana, elucidated the genetic code.

Contd...

Year	*Event*
1967	Gilbert discovered DNA ligase, the enzyme used to join DNA fragments.
1972-1973	DNA cloning techniques were developed by Boyer, Cohen and Berg and their colleagues.
1975	EM Southern developed the technique of Southern blotting involving gel transfer and hybridization for detection of specific DNA sequences.
1975-1977	Sanger & Barrell and Maxam & Gilbert developed rapid DNA sequencing methods.
1981-82	Palmiter and Brinster produced transgenic mice; Spradling and Rubin produced transgenic fruitflies and J. Schell produced transgenic plants.
1985	K. Mullis and coworkers invented polymerase chain reaction (PCR).
1983	A Murray, J Szostak constructed the first artificial chromosome.
1984	Alec Jeffreys developed the technique of 'DNA fingerprinting'.
1986	H.R. Horvitz discovered programmed cell death (PGD).
1986	'Human Genome Project' conceptualized
1987	First comprehensive human genetic map produced.
1987	K.A Barton isolated Bt gene from *Bacillus thuringiensis*.
1988	'National Centre for Human Genome Research' (NCHGR) created.
1990	'Human Genome Project' (HGP) formally launched.

Contd...

Year	*Event*
1990	W.F. Anderson conducted first 'gene therapy' experiment.
1990	J.G.K. Williams et al. developed the technique of RAPD.
1992	Human genetic map for all 23 chromosomes developed.
1995	TIGR published the complete sequence of the genome of a living organism *(Haemophilia influenzae).*
1995	M. Schena developed DNA chip "microarray" system.
1995	P. Vos developed AFLP technique.
1996	First eukaryotic genome (budding yeast) sequenced.
1996	Calgene released first transgenic crop (flavr *savr tomato)* for commercial cultivation.
1997	I Wilmut et al. cloned first animal, a sheep earned 'Dolly'.
1998	Genome of a multicellular organism (C. *elegans)* fully sequenced.
1998	'International Rice Genome Sequencing Project' (IRGSP) launched.
2000	*Drosophila* genome sequence published.
2001	Working drafts of human genome sequence published.
2002	Syngenta & BG1 published draft rice genome sequences.
2002	Draft mouse genome sequence published.
2004	Rat genome sequence published.

Selected Nobel Prizes (in Physiology/Medicine or Chemistry) Awarded for the Work Relevant to Molecular Biology and Genetic Engineering

2003	P.C. Lauterbur and P Mansfield	Magnetic Resonance imaging
2002	S. Brenner, HR Horvitz and J.E. Sulstonorgan	Genetic regulation of development. and programmed cell death.
2001	L.H. Hartwell, R Hunt and PM Nurse	Key regulators of the cell cycle.
2000	A. Carlsson, P.Greengard and E Kandel	Signal transduction in the nervous system.
1999	Gunter Blobel	Signal sequences and intracellular protein trafficking.
1998	RF Furchgott, LJ Ignarro, F Murad	Nitric oxide (NO) as a signaling molecule.
1997	SB. Prusiner	Discovery of prions (infectious particles).
1996	PC Doherty and RM Zinkernagel	Cell mediated immune defence and 'MHC restriction".
1995	EB Lewis, C Nusslein-Volhard,	Genetic control of embryo development in *Drosophila* and EF Wieschaus.
1994	M Rodbell and AG Gilman	Discovery of G-proteins.
1993	K Mullis, M. Smith	Polymerase chain reaction; site directed mutagenesis.
1993	PA Sharp and RJ Roberts	Discovery of split genes.
1992	EG Krebs and EH Fisher	Reversible protein phosphorylation as a regulatory mechanism.
1991	E Neher and B Sakmann	Function of ion channels in cells.
1988	JM Bishop and HE Varmus	Cellular origin of retroviral oncogenes.
1987	S Tonegawa	Genetic basis of antibody diversity.
1984	NK. Jerne, GJF Kohler and C Milstein	Hybridoma and monoclonal. antibodies.

Contd...

1982	A Klug	Three dimensional structure of tRNA.
1983	Barbara McClintock	Discovery of mobile genetic elements.
1980	P Berg, W Gilbert, F Sanger	Determination of base sequences in nucleic acids.
1978	W Arber, D Nathans, H Smith	Discovery and application of restriction enzymes.
1975	D Baltimore, R Dulbecco, HM Temin	Interaction of tumour viruses with cell.
1974	A Claude, C de Duve, G Palade	Structural/functional organization of the cell.
1970	A. Kossel	Nucleic acid contains four nitrogen containing compounds (A,T,C,G).
1969	M Delbrnuck, AD Hershey, SE Luria	Replication and structure of viruses
1968	R Holley HG Khorana, MW Nirenberg	Sequencing of the tRNA; first interpretation of the genetic code and its function in protein synthesis.
1965	F Jacob, A. Lwoff and J. Monod	Genetic control of enzyme and virus synthesis.
1964	K Bloch and F Lynen	Cholesterol and fatty acid metabolism.
1962	MF Perutz, JC Kendrew	Determination of the TD structure of hemoglobin and myoglobin.
1962	FHC Crick, JD Watson, M. Wilkins	Double helix structure of DNA.
1959	S Ochoa and A Kornberg	Biological synthesis of RNA and DNA.
1958	GW Beadle, EL Tatum, and J Lederberg	Gene action through enzymes; recombination and organization of genetic material in bacteria.
1958	F Sanger	Determination of amino acid sequence of insulin.

Contd...

1955	V du Vigneaud	Synthesis of the first protein (oxytocin) in the laboratory.
1955	AHT Theorell	Nature and mode of action of exidative enzymes.
1953	H Krebs FA Lipmann	Discovery of citric acid cycle; discovery of co-enzyme A.
1945	A Fleming, EB Chain, HW Florey	Discovery of Penicillin.

***Definition and History of Molecular Biology* :** The history of molecular biology can be traced back to the origin of organic chemistry, when for the first time in 1807, the Swedish Chemist J.J. Berzelius classified matter into two types of molecules; the combustible substances (derived directly or indirectly from living organisms) were called "organic" and the rest of the substances constituting matter were described as "inorganic".

Ever since this distinction was made between organic and inorganic substances, for almost a century and a half, significant progress was made in the field of organic chemistry. However, this progress in chemistry was without any understanding of the significance of these molecules in determining the structure and function of living systems.

During the study of living matter, it was only in the nineteenth century, that chemists also identified a large group of substances, which on heating changed from liquid to a solid state. Substances like white of egg, casein in milk and globulin in blood belonged to this category of organic substances. These substances were studied by Dutch chemist, G. J. Mulder, who on suggestion of Berzelius, called them "protein" (meaning '*of first importance*').

Later, these substances called "proteins" were found to be the key molecules of life, and were found to be more important than other substances like carbohydrates and fats found in the living cells. By the end of the nineteenth century, biochemists also became certain that proteins are built up of the building blocks called amino acids.

During early years of nineteenth century, Emil Fischer also demonstrated that amino acids can be condensed to give bigger molecules with elimination of water. Fisher also used the word "peptide" for small synthetic chains derived from condensation of amino acids and "peptide link" for the link between carboxyl group of one amino acid with the amino group of another amino acid.

While proteins were being studied in the nineteenth century, in 1869 a Swiss biochemist, named Friedrich Miescher, while breaking down the protein of cells with pepsin (an enzyme, which is specific for breaking down proteins), discovered that pepsin did not break the cell nucleus. He later discovered that the nucleus largely consisted of a phosphorus containing substance, which did not resemble proteins and was called 'nuclein' as against 'protein'.

Twenty years later, when 'nuclein' was found to be strongly acidic, it was renamed 'nucleic acid', which on hydrolysis gave four types of nitrogen-containing compounds (collectively called bases) named 'adenine', 'guanine', 'cytosine' and 'thymine' (discovered by German biochemist, A. Kossel, who received Nobel Prize in Medicine and Physiology in 1970).

In 1911, P.A.T. Levene also discovered that nucleic acids contained sugar molecules each with a five carbon ring. He also found for the first time that nucleic acids could be of two types, deoxyribonucleic acid (DNA) containing deoxyribose sugar and ribonucleic acid (RNA) containing ribose sugar.

As shown above, it was obvious by 1940s, that protein and nucleic acids were two important and major classes of molecules that are found in living cells, although other molecules described as carbohydrates, fats and lipids, etc. were also found. It also became known that DNA content in individual cells of a particular living organism was constant, while the protein could vary.

However, proteins and nucleic acids were closely associated to form, what were later described as "nucleoproteins". Therefore, at that time a debate was on to find out, which of the two molecules, the 'protein' or the 'nucleic acid' constituted the hereditary material. It was only in 1944, that three American biochemists (O.T. Avery, CM. Macleod and M. McCarthy) jointly discovered for the first time that DNA is the genetic material.

In UK, later in 1953, J.W. Watson and F.H.C. Crick resolved the double helix structure of nucleic acids, on the basis of results of X-ray crystallography of nucleic acid molecules obtained by two other scientists, R. Franklin and M. Wilkins. For this work, Watson, Crick and Wilkins received the 1962 Nobel Prize in Medicine/ Physiology. Recently, in the year 2003, fifty years of this important work was celebrated all over the world.

As shown above, by the middle of the twentieth century, it was fairly clear that the key molecules, which form the basis of all life are 'proteins' and 'nucleic acids'. The study of these molecules laid the foundation of the new discipline, called 'molecular biology', which not only wiped out the borderline between the living and the non-living, but also revolutionized all areas of biology.

Definition and History of Genetic Engineering (Recombinant DNA) : Genetic engineering is defined as "the technique of cutting foreign DNA at specified places and inserting the cleaved foreign DNA fragment thus obtained into another DNA molecule (called vector) similarly cut". The product obtained by inserting foreign DNA into another DNA (vector) is described as 'recombinant DNA'. If this recombinant DNA is used for transformation of host cells (e.g. *E. coli* or yeast cells or human cells in culture) and allowed to multiply in these host cells along with foreign DNA, one can obtain millions of copies of foreign inserted DNA.

The technique is then described as 'DNA cloning'. Similar amplification of DNA can also be obtained through the technique of polymerase chain reaction (PCR). The PCR technique was developed in 1985, by Kary Mullis, who received the 1993 Nobel Prize in Chemistry, jointly with Michael Smith for his discovery of site directed mutagenesis Genetic engineering of an organism, however, involves introduction of DNA, genetically manipulated *in vitro,* into living cells, so that this DNA becomes a part of the hereditary material of living cells.

Since this technique involves manipulation and introduction of foreign unrelated DNA into cells of living organisms, the technique of genetic engineering differs from the technique of "transformation". Transformation leading to phenotypic modification, involves transfer of DNA from a different genotype of the same species. This was done for the first time by F. Griffith (1928), and used later by Avery et al. (1944) to demonstrate for the first time in pneumococcus that DNA is involved in transformation.

Until the early 1970s, DNA was the most difficult cellular molecule to be analysed. However, the situation changed dramatically during 1970s due to development of a series of techniques, collectively described as 'recombinant DNA technology'.

The most important of these techniques include the following : (i) *cleavage of DNA,* at specific restriction sites by restriction enzymes; *(ii) DNA cloning,* where a single DNA molecule can be copied to generate billions of identical

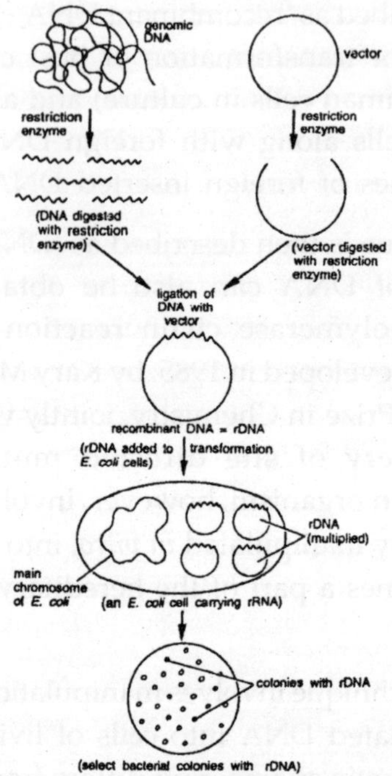

Different steps involved in 'genetic engineering' or 'gene cloning'.

molecules; (iii) *rapid sequencing of nucleotides* in a purified DNA fragment; *(iv) nucleic acid hybridization*, permitting identification of a specific sequence of DNA or RNA; (v) *DNA engineering*, by which DNA sequences are altered to make modified versions of genes, which are inserted back into cells or organisms; (vi) *polymerase chain reaction (PCR)*, for faithful amplification of a DNA segment into millions of copies within a few hours; (vii) *whole genome sequencing*, using capillary based automatic sequencing machines making it possible to sequence even large whole genomes of eukaryotes.

Some major steps leading to the development of molecular biology and recombinant DNA technology are summarized in Table. The steps involved in 'genetic engineering', also described as 'gene cloning' are shown in Figure. It can be seen in this figure that following of the above techniques are involved in gene cloning : (i) a method of DNA cleavage at specified places; (ii) a method of joining (ligating) DNA fragments to yield chimaeric DNA; (iii) a method of transformation to infect bacterial cells with chimaeric DNA; and (iv) a method for selecting transformed bacterial cells containing chimaeric DNA.

Scope of Molecular Biology and Genetic Engineering : Molecular biology and genetic engineering have become so pervasive in all areas of biology, that a study in any subject of biology is considered incomplete, if it is not studied at the molecular level using techniques of recombinant DNA. For instance, using the techniques of reverse genetics, genes for a variety of morphological traits are being identified and the mechanisms underlying the development of these traits are being uncovered. Plant physiology and animal physiology are also being studied at the molecular level.

Structures and functions of different cell constituents are being resolved at the atomic level, giving a new dimension to the study of cytology, which is more appropriately described as cell biology now. For example cell wall, cell membrane, cytoskeleton, nuclear membrane, chromosomes, nucleolus, different cell organelles (chloroplasts, mitochondria, peroxisomes, ER, Golgi, etc.), have been studied achieving very high resolution sometimes approaching atomic level. These developments gave birth to new areas of biology like 'genomics', 'in silico biology', 'systems biology', 'integrative biology', 'bioinformatics', etc.

Genomics and Proteomics

In recent years, due to the possibility of sequencing whole genomes of microbes as well as higher plants and animals, attempts are being made to study all genes in an organism. DNA sequences are being deposited in databases and are being annotated to resolve not only the function, controlled by each individual gene, but also the gene × gene, protein × protein and DNA × protein interactions.

Genomics research also prompted the study of all proteins within a cell, and the subject is described as "proteomics", which is more powerful than even genomics. Transcriptomics (the study of all mRNA within a cell) and metabolomics (study of all metabolites within a cell) are the other offshoots of the science of genomics, which is being very actively pursued all over the world. Enormous data has been generated through research in this area, and utilization of this data through the use of computers gave birth to 'in silico biology' and 'bioinformatics'.

Silico Biology and Bioinformatics

In recent years, science of biology has witnessed a revolution due to extensive developments in the application of computers in biology (particularly the software developments) and due to the accumulation of extensive data in the databases. A large amount of literature is also freely available on computer, so that the access to information and knowledge has become much simpler, easier and cheaper than ever before. Research can now be conducted and knowledge generated by mere use of computers, where one can play with the freely availably data and many softwares. 'Computer modelling' of molecules, networks and metabolic pathways has been possible and experiments could be designed to test these models. This area of science has been described as 'in silico biology' and developments in the field of bioinformatics made this area of research highly rewarding. In view of these developments, it has now become necessary for every biologist to be a bit of bioinformatician.

Systems Biology and Integrative Biology

Since data on morphology, physiology, biochemistry and genetics of an organism as a whole is now available, there is a need to study a process as a whole in all its aspects. This has given birth to the subject popularly described as 'systems biology' and 'integrative biology'.

In this new science, one needs to integrate information from all areas (morphology, physiology, ecology, genetics, biochemistry, etc.) to draw a picture of the entire system in an integrated manner. It was not possible earlier, because experimental tools were available for the study of one aspect at a time and that too in a fragmented manner. For instance, one could study the variation in morphology, without knowing the genetic basis and the underlying processes responsible for this variation Genetic control of a single trait could be resolved, without knowing the molecular mechanism through which the genetic control operates generating the variation, which we observe. The boundaries of different branches of biology have disappeared now.

Interdependence of 'Molecular Biology' and 'Genetic Engineering' : It has also been shown earlier in this chapter that the developments in molecular biology gave birth to the branch of 'genetic engineering' in 1970s. However, ever since its birth, genetic engineering has also been utilized for generating knowledge in the area of molecular biology.

Many riddles of molecular biology could be resolved, because new experiments could be designed to discover phenomena and concepts relevant to biology. For instance, mechanisms of DNA replication, protein synthesis and its regulation, and that of differentiation of cells and development of body parts could be studied using the techniques of genetic engineering. Therefore, molecular biology and genetic engineering are not the two independent areas of biology, but are interactive, one benefitted by the other in generating knowledge.

4

Developmental Genetics

The genetic control of phenotypic traits at different levels including phenotypic level, biochemical level, and molecular level. This has helped us in developing an understanding of how an individual gene consisting of a DNA segment may function in giving rise to a phenotype mediated through the synthesis of a protein and how can the activity of such a gene be regulated. However, the coordinated activity of different genes and the regulation of this activity lead to different developmental patterns, starting from an egg or a zygote to the fully developed adult. This aspect is most difficult to study due to the limitation, which our experimental designs impose on such a study. This subject is sometimes described as the study of differentiation and can be studied, again at various levels including morphology, biochemistry, molecular biology and genetics.

Only a coordinated study involving these different approaches will allow us to develop a proper understanding of how and why a particular pattern of development is followed by an individual. As an example, we may consider the development of an adult human (with about 10^{14} cells) from a single cell *i.e.* zygote derived due to fusion of an egg and a sperm. A popular question in this connection is : how does a ball of seemingly identical cells (the blastula), derived by mitotic divisions from single fertilized egg, becomes differentiated into an organism of recognizably different regions and tissues?

The process of differentiation is preceded by cell specification and cell determination as parts of cell commitment. This determines the ultimate fates of individual cells that are seemingly identical, but differ in their cellular determinants. Three types of specifications were discussed, but in each case, once the cells are committed, they differ in a very specific manner. From this initial difference in cells, during the process of development, anterio-posterior and dorsoventral distinctions will arise. A vast and amazing array of cell types will be found in the mature individual, so that even two adjacent cells may have different phenotypes and functions.

The developmental pathway is so precise, that each step of development is dependent on the preceding stage. Any disruption in the normal sequence of events results in an abnormal phenotype, sometimes due to mutations. It is these abnormal phenotypes and the gene mutations responsible for them, which help in the proper understanding of the sequence of events and the basis for this sequence.

In any study of development, we know that every cell of an organism, at any stage of its development, contains the entire genome with all genes. Then, how these genes regulate their activity in a programmed manner to confer the complexity observed in the multicellular organism? An analysis of this problem at the genetic level requires a simple system amenable to experimentation, in order to probe into the mechanism involved. In this chapter, an effort will be made, with the use of some examples, to demonstrate how in recent years some information on the genetics of development became available. We will also discuss, what kind of questions are likely to be answered in future, due to the powerful techniques of recombinant DNA developed during 1970s and more particularly during 1980s and 1990s.

Factors Controlling Development

While the genetic constitution of the zygote largely controls the course of development, there are several other factors, which have been shown to play an important role in controlling the developmental patterns. These include the following :

(i) Cytoplasmic factors in some cases play a very important role in early embryonic development. For instance, yolk and mitochondria are not evenly distributed in an amphibian egg, and get unevenly partitioned into different daughter cells. Experimental demonstration of the extranuclear control of development has been possible by grafting a part of cell cortex (without nuclei) from an embryo into a fertilized egg, which led to the development of an extra spinal cord.

(ii) Environmental factors also play an important role in development as shown in phenocopies in an organism, where same genetic constitution leads to different phenotypes if grown in different environments. However, treatment to a developing embryo with a definite chemical or temperature, etc., should be given at a specific critical period during development in order to obtain each particular phenocopy.

(iii) Maternal effects on development have been witnessed in many cases, where the cytoplasm of the egg was found to, regulate the early events in embryonic development. For instance, even an enucleated egg of a frog will still undergo the normal early cleavage, although one may argue that the architecture and composition of the cytoplasm are themselves under the control of the nuclear genome of the mother.

Changes in the Nucleus During Development (Nuclear Transplantation) : We often argue that all cells of an organism are identical in its genetic content and that the development proceeds by differential expression of these genes. The totipotent nature of differentiated cells in many organisms (particularly plants) is often put forward as an evidence suggesting that all cells are similar. Cytological examination and DNA content analysis also, in most cases, revealed that

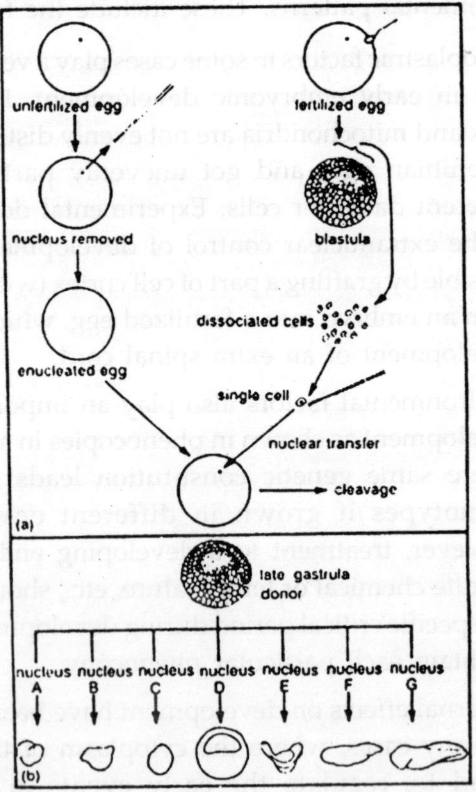

(a) Technique of transplantation of a nucleus from a blastula stage into an enucleated stage; (b) results of transplantation of several gastrula nuclei in the frog Rana.

same chromosome number and same DNA contents (barring exceptions like antibody producing cells, etc.) are found in all cells. However, nuclear transplantation experiments (Fig.) suggested that in frog *(Rana),* the somatic nuclei from the blastula cells and those from the gastrula stage each differ among themselves (due to epigenetic changes), so that after gastrulation the nuclei differentiate irreversibly. However, in contrast to this situation in *Rana,* the situation in *Xenopus* differs, where the nuclei of both blastula as well as gastrula are totipotent and, when transplanted in eggs, can give rise to normal adults. It has, however, been shown that the differentiated cells from a variety of tissues do differ qualitatively. One of these differences involve the degree of methylation or demethylation of DNA representing different genes, a phenomenon known to be involved in the regulation of gene expression.

Gene Regulation and Use of Markers During Development: Another aspect of development and differentiation is the differential activation of various genes at different stages of development. Such a differential activation of developmental genes has actually been demonstrated in some cases. In recent years, mutations in specific genes have been identified which lead to failure of the development of specific organs. Such mutants have been studied in large number and have been very useful in identification, isolation and cloning of specific genes involved in developmental pathways. Morphological and genetic markers have also been used for the study of development, particularly for the preparation of 'fate maps' of early embryos.

Gene Regulation During Early Embryonic Development : It has been demonstrated in several organisms (including *Drosophila, Xenopus* and other amphibians, mouse, etc.), that just after fertilization, cell divisions are rapid and are controlled by maternally derived information in the cytoplasm. The zygotic nuclei do not exercise any genetic control in these early embryonic cell divisions, but start playing a role only after initial cell divisions are over. Techniques have now been developed which enable to locate the stage in development, at which the embryonic (or zygotic) nuclei start playing a role in the development. These approaches include the following:

Nuclear Transplantation : Nuclear transplantation from sperms of one species into enucleated eggs of another species has been utilized to locate the stage at which paternal influence becomes obvious.

Location of Stage for RNA Synthesis : The stage at which some gene product (either RNA or protein): encoded by paternal genes becomes available, can be located by using drugs like actinomycin D, which would inhibit RNA synthesis on DNA template, causing lethality. The stage at which actinomycin D starts having lethal effect will suggest that the products of transcription or translation are essential for survival of the developing embryo. In amphibians, zygotic genes start functioning at the time of gastrulation. In mouse embryos, however, lethal effects of actinomycin D begin to appear around 9 or 16 cell stage of development.

Activation of Specific Genes in Development : There are a number of genes, which act early in development and whose products are found in large quantities in differentiated cells thus facilitating their identification. In *Xenopus*, until gastrulation (when differentiation of three similar cell layers occurs), maternal mRNA is found in good quantity and no new genes are activated. From gastrulae, mRNA could be isolated on poly-dT column, due to the presence of poly-A tails. Such mRNA contains both the maternal mRNA and zygotic mRNA and has been used for synthesis of cDNA through the use of reverse transcription. This cDNA has been used for differential or subtractive hybridization with mRNA from unfertilized egg, so that cDNA not hybridized with this mRNA could be isolated as single stranded DNA and used for making double stranded cDNA. This double stranded cDNA has been used to make a library of up to 1,50,000 clones (many of them will be present repetitively in this library). This library should represent DNA sequences complementary to mRNA differentially synthesized at the gastrulation stage.

When a number of these clones bound on filters were hybridized with radioactively labelled mRNA extracted from embryos at different stages of development, one could identify the number of genes and the degree of their expression at different-developmental stages. Such experiments led to the following conclusions :

(i) The cDNA clones represent genes that are activated during gastrulation.

(ii) Level of expression of different genes during gastrulation varies from gene to gene.

(iii) Most of these individual mRNAs declined in abundance in later stages, although total mass of mRNA increases, suggesting that most of these genes are shut off by the tadpole stage.

Puffing Patterns in Giant Chromosomes : The giant polytene chromosomes or salivary gland chromosomes of Diptera, is a system which has been utilized to demonstrate the visible differential gene activity. The swollen puffs and *Balbiani rings* are the sites of RNA synthesis as demonstrated by incorporation of H^3 labeled uracil at these sites, after incubation of larvae in this medium. The size and patterns of these puffs differ at different stages of development and indicate predictable differential gene activity. This has been demonstrated by giving heat shocks or treatment with steroid hormones, which lead to correlated changes in puffing patterns on the one hand and that of the appearance of specific mRNA and proteins on the other.

Use of Markers in Developmental Genetics : Both phenotypic and genetic markers have been utilized in studies of development. For instance, by marking specific cells of an amphibian embryo, with visible carbon particles, movements of marked cells and their descendants during cell divisions could be traced during cell divisions. Using this technique, fate maps could be prepared, showing the destinies of the descendants of particular cells in early stages of embryogenesis. Similarly, one may mark the cells genetically, by taking them from different sources, fuse them and then trace their fate during development.

Tetraparental Mouse : Two mouse embryos (at 8 or 16-celled stage) each derived from two parents, can be fused together, after stripping off their surrounding membranes (using pronase enzyme). This may then be implanted in a host female, so that a tetraparental embryo will develop as a single mosaic individual. As an example, Beatrice Mintz (a lady scientist) fused embryos (8-16 cells) from two strains, one having black fur and the other having white fur. The mosaic individual which developed, had bands of white and black fur circling the whole body. This suggested that the cells, whose descendants produced pigment, move after fusion of embryos and get lined up randomly in pairs, each member of a pair dividing to form a clonal sheet extending halfway around the body to meet the other sheet. Utilizing tetraparental mice, Mintz also resolved the controversy whether a multinucleate muscle cell originates due to a series of nuclear divisions without the division of cytoplasm or due to successive fusion of different myoblasts. Using genotypes with different forms, *Id-1^a* and *ld-l^b* of an enzyme *(isocitrate dehydrogenase)*, it could be shown that a striated muscle cell is formed by myoblast fusion, as evident from the formation of hybrid band (dimer) of the enzyme.

Genetic Mosaics in Drosophila : In *Drosophila* also, genetic mosaics have been produced as a result of two different phenomena, namely mitotic crossing over and chromosome loss. Mitotic crossing over leads to the production of twin spots of homozygous tissues and size of these spots depends on the stage at which mitotic crossing over occurred (earlier the crossing over, bigger is the size of spots). A study of a set of twin spots of different sizes would help to trace the cell lineage of an organ or a part of it. Atleast two cases are worth mentioning :

(i) A heterozygote w^{co}/w (w = white eye) gave twin spots, which allowed tracing of cell lineage of lower half of the eye to eight early larval cells,

(ii) A study of twin spots in a heterozygote, *yHw+/++ sn*^{3} (y = yellow body; *Hw* = extra bristles along wing veins; *sn*^{3} = short gnarled bristles) gave evidence that Hw allele acts before the last 12 hours of larval life to imprint its phenotype irreversibly on the cell committed to form a wing.

A number of mutants in *Drosophila* are now known, which have a tendency for chromosome loss giving rise to two types of tissues one with full complement of chromosomes and the

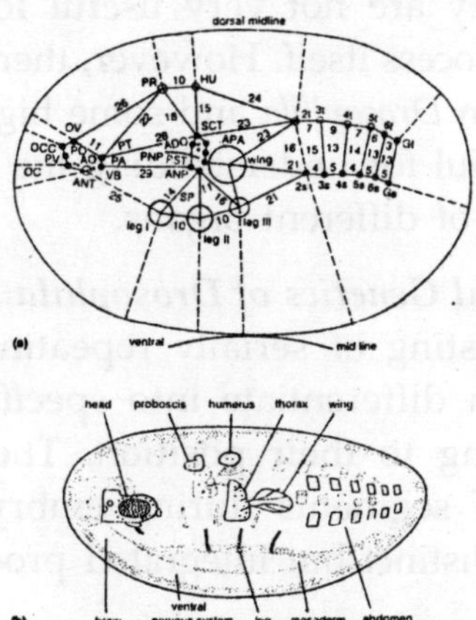

(a) An embryonic fate map of adult structures in Drosophila; the map shows foci for various structures and individual bristles; the distances are shown in map units called sturts. (b) A redrawn version of the above fate map, showing that the arrangement of structures in adult retains the pattern shown in the embryonic fate map (redrawn from Benzer, 1973).

other with the deficiency of a specific chromosome. These tissues indicate that cells more closely related by cell division tend to stay together and therefore have been used for preparing fate maps in *Drosophila* (Fig.). Such fate maps have relationship with the actual spatial distribution of determined cells in the blastoderm.

Genetics of Development

The embryologists initially studied the developmental process, by destroying, removing, replacing or shifting specific parts of an embryo followed by examining the consequences. The same objective can be achieved more conveniently by a geneticist through the study of developmental mutants, characterized by disruption of developmental sequences. Although a number of lethal mutations have been identified in a variety of plants and animals, they only suggest the presence of genes or chromosome segments (some lethal mutations are actually deficiencies for small chromosome segments), which are essential for survival.

However, they are not very useful for a study of the developmental process itself. However, there are a number of other mutations in *Drosophila* and some higher plants, which proved to be useful for understanding the genetic control of the development of different organs.

Developmental Genetics of Drosophila : Insect bodies are metameric, consisting of serially repeating units (or body segments), which differentiate into specific structures and patterns according to their position. The generation and diversification of segments during embryo organogenesis depend on two distinct but integrated processes:

(i) the subdivision of the embryo into reiterated units, and

(ii) the specification of their pathway to differentiation. In many insects, segments are generated sequentially as the cells of the embryo proliferate.

By contrast, in *Drosophila* and other so-called long germ-band insects, the entire body plan is established simultaneously at the blastoderm stage. The adult fly consists of (i) head, (ii) three thoracic segments (prothorax, mesothorax and metathorax, designated as Tl, T2, T3) and (iii) eight abdominals (AB1 to AB8) as shown in Figure. The T2 carries the wing and the second leg and T3 carries the first leg. These segments can be seen even in the larval stage.

Genes controlling, development in *Drosophila* can be broadly classified into three groups:

(i) genes controlling anterio-posterior polarity;

(ii) genes controlling dorso-ventral polarity and

(iii) genes that respond to coördinates established by anterio-posterior and dorso-ventral axes and are thus expressed due to specific positions of organs primordia.

The pioneering work of the genetic control of embryonic development and differentiation in *Drosophila* was so important in the area of developmental genetics that the Nobel Prize in Medicine for the year 1995 was awarded to the following three scientists for this work: Edward B Lewis (USA), Christianne Nusslein-Volhard (Germany) and Eric Wieschaus (USA).

Gradient Model and Organizing Centres : It has been shown that the cells and their genes can respond to gradients of soluble substances (morphogens) in a very specific manner, with the help of enhancer sequences associated with these genes (Fig.). These gradients in their turn are also under genetic control. Following three sets of genes are known, which control antero-posterior polarity:

(i) the genes encoding gradient morphogens for the anterior part of the embryo (*e.g. bicoid* or *bed* gene);

(ii) the genes that are responsible for organizing posterior or abdominal region (*e.g. nanos* or *nos)* and

(iii) the genes encoding proteins that produced the terminal regions (*e.g. torso or tor*) (Fig.). The genes belonging to these three categories along with the details of mutant phenotypes and proposed functions, are listed in Table.

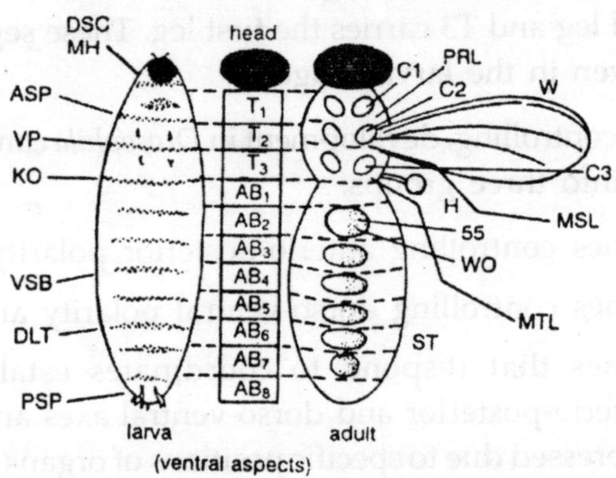

Ventral view of body segments of Drosophila larva and adult stages (note that each segment found in adult can be identified in the larva).

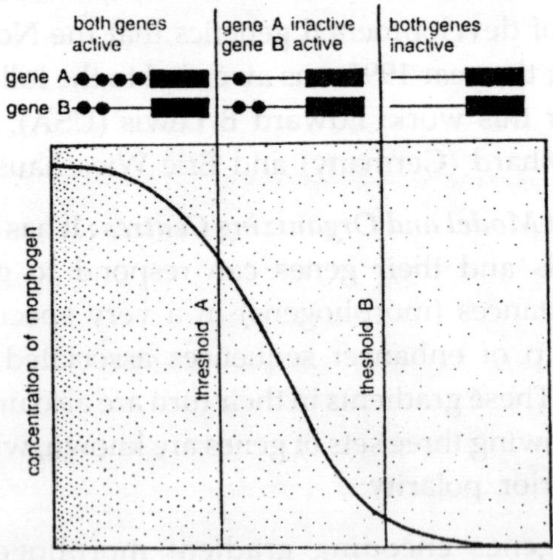

A model for morphogen gradients establishing positional information; assumed receptors for morphogens are enhancer elements of two genes (A, B) controlling cell fate, but the receptors can be located elsewhere also, (redrawn from Gilbert 'Developmental Biology', 1994).

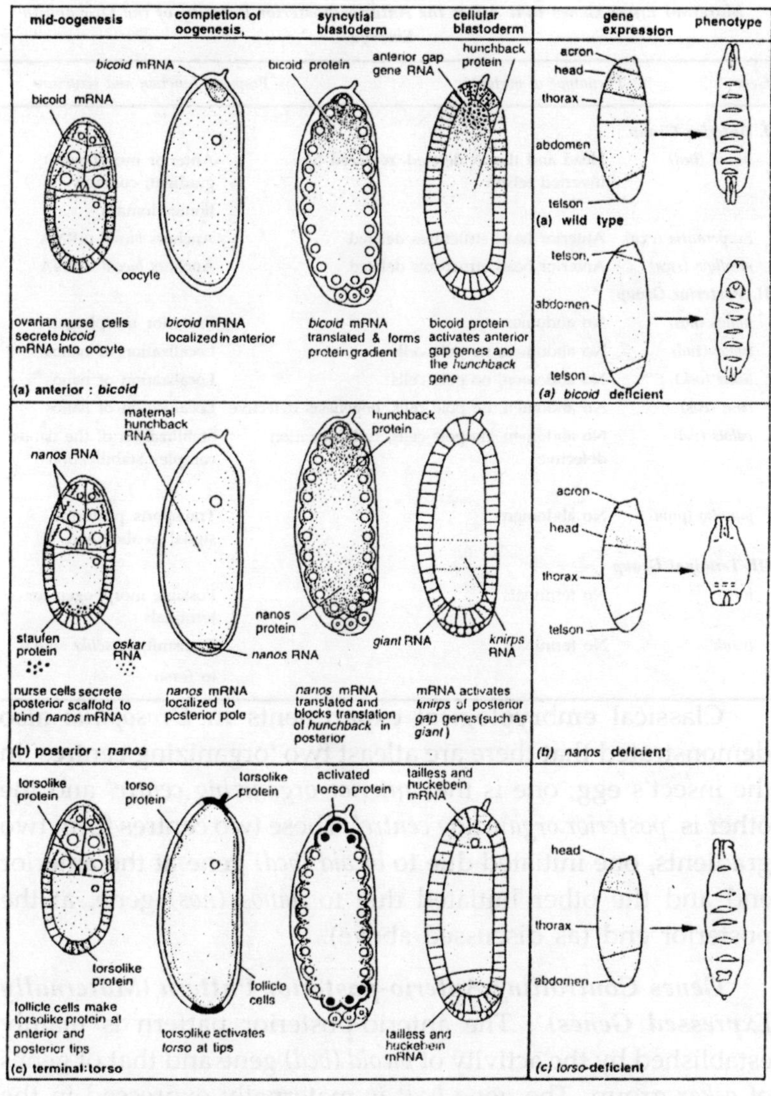

Three independent genetic pathways, which interact to form the anterio-posterior axis of Drosophila embryo; the patterns are established when the localized maternal products[1] activate or repress specific genes in different regions of the embryo (redrawn from Gilbert 'Developmental Biology, 1994).

Maternal Effect Genes that Affect the Anterior-posterior Polarity of the Drosophila Embryo

Gene	*Phenotype of mutants*	*Proposed function and structure*
I. Anterior Group		
bicoid (bcd)	Head and thorax deleted, replaced by inverted telson	Anterior morphogen gradient; contains homeodomain
exuperantia (exu)	Anterior head structures deleted	Anchors *bicoid* mRNA
swallow (swa)	Anterior head structures deleted	Anchors *bicoid* mRNA
II. Posterior Group		
nanos (nos)	No abdomen	Posterior morphogen
tudor (tud)	No abdomen, no pole cells	Localization of nanos
oskar (osk)	No abdomen, no pole cells	Localization of nanos
vasa (vas)	No abdomen, no pole cells, oogenesis defective	Localization of nanos
valois (val)	No abdomen, no pole cells, cellularization defective	Stabilization of the nanos complex stabilization
pumilio (pum)	No abdomen	Transports posterior signal to abdomen
III. Terminal Group		
torso	No terminals	Possible morphogen for terminals
trunk	No terminals	Transmits *torsolike* signal to torso

Classical embryological experiments in *Drosophila* also demonstrated that there are atleast two 'organizing centres' in the insect's egg; one is the *'anterior organizing centre'* and the other is *'posterior organizing centre'*. These two centres form two gradients, one initiated due to *bicoid (bcd)* gene at the anterior end and the other initiated due to *nanos (nos)* gene, at the posterior end (as discussed above).

Genes Controlling Anterio-posterior Pattern (Maternally Expressed Genes) : The anterio-posterior pattern is mainly established by the activity of *bicoid (bcd)* gene and that of genes of *oskar* group. The gene *bcd*$^{+}$ is maternally expressed in the ovary in specialized cells, which form a cluster around the future anterior pole of the developing oocyte. As *bcd* RNA enters oocyte, it is trapped, apparently by components of cytoskeleton encoded by the genes *swallow (swa)* and *exuperantia (exu)*, and thus becomes localized at the anterior pole of the oocyte. The *bicoid* mRNA is though localized thus in the anterior portion of the oocyte, its product on translation forms a gradient, with the highest concentration in the anterior of the egg. The mutants for genes *exuperantia (exu)* and *swallow (swa)* have phenotypes similar to that for *bicoid* deficient embryos, but the effects are less severe. These mutant embryos lack anterior structures and have an extended mouth and the thoracic region.

Similarly, and at about the same time when *bed* is expressed maternally at the anterior end of the egg, the products of genes *(nanos = nos* gene) of *oskar* group get deposited and localized at the posterior end of the egg. The products of several genes *(oskar, valois, vasa, staufen* and *tudor)* are needed to place the *nanos* mRNA into the posterior part of the egg. Thus the egg gets polarized at the morphological level as well as at the molecular level. There is a third set of genes (*e.g. torso = tor*), whose products are localized at both the tips of the embryo and are thus involved in the production of the terminal regions

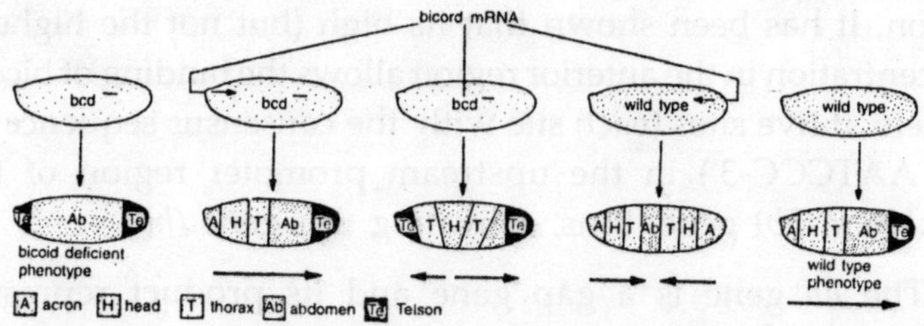

Experiments, where bicoid-deficient and wild* Drosophila *embryos are injected with biocid mRNA, showing the formation of head structures at the point of injection (redrawn from Gilbert* 'Developmental Biology', *1994).

at both ends of the embryo. The gradients established due to above three sets of genes and the mutant phenotypes are shown in Figure, and the details of these genes are listed in Table. Mutants or deficiency for *bed* gene produce eggs that develop into embryos with no head or thorax, but with abdomen flanked by telson at both ends.

The removal of products from anterior pole leads to similar results and injection of mRNA at the appropriate region of each embryo, caused these embryos to have anterio-posterior polarity (Fig.). Further, any location, where the *bicoid* mRNA was injected became the head.

If the mRNA was injected in the posterior region, an embryo with two heads developed. Similarly, mutants for any *osk* group gene develop normal head and thoracic segments but lack entire abdomen. The mutants for *tor* or for any other gene of the group lack 'acron' and 'telson' at the two ends and possess only head, thorax and abdomen (Fig.).

After fertilization, *bcd* RNA moves backward and *osk* RNA moves forward to about the middle of the zygote, each establishing a *gradient*. Subsequently, these maternally derived products bind with DNA and thus interact with the zygotic genome.

We have seen above that a gradient for bicoid protein is established, with the highest concentration in the anterior region. It has been shown that its high (but not the highest) concentration in the anterior region allows the binding of bicoid protein at five sites (each site with 'the consensus sequence 5'-TCTAATCCC-3') in the upstream promoter region of the *hunchback (hb)* gene, thus, activating this gene *(hb).*

The *hb* gene is a gap gene and its product represses abdominal-specific genes, thus permitting the formation of head and thorax rather than abdomen in the anterior region (for gap genes, see below). At a very high concentration (at the very anterior tip), 'bed' protein can also bind and activate head-specific gap genes whose products are needed to specify the anterior region of the head. These head-specific gap genes include *buttonhead (btd), empty spiracles (ems)* and *orthodenticle (otd).*

The action of *nanos* mRNA in the posterior region is same as that of *bicoid* in the anterior region. For instance, the protein product of *nanos* gene (described as posterior morphogen) represses the translation of *hunchback* mRNA, which is present throughout the embryo (although in the anterior region, more *hb* mRNA can be made due to binding of 'bed' protein at the promoter enhancer sites associated with *hb* gene; (Fig.).

In this manner, bed and nanos proteins help in the (c) concentration of hunchback protein specifying positions of transcription of three gap genes establishment of a gradient for

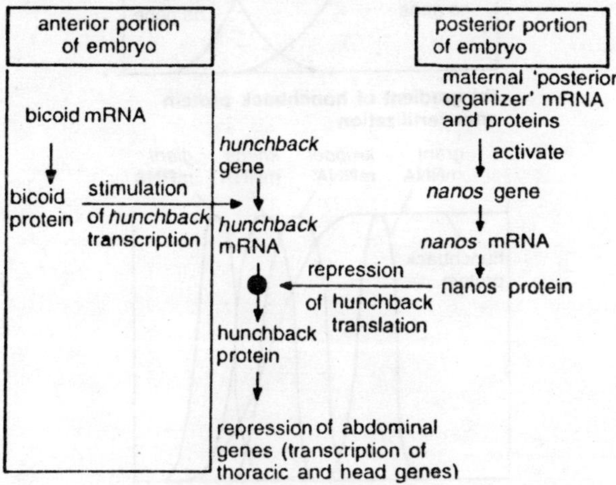

A model for the action of bicoid and nanos proteins in specifying the anterio-posterior axis of the embryo; bicoid protein activates hunchback transcription and translation in the absence of nanos protein, which represses hunchback translation in the posterior region. (redrawn from Gilbert 'Developmental Biology', 1994).

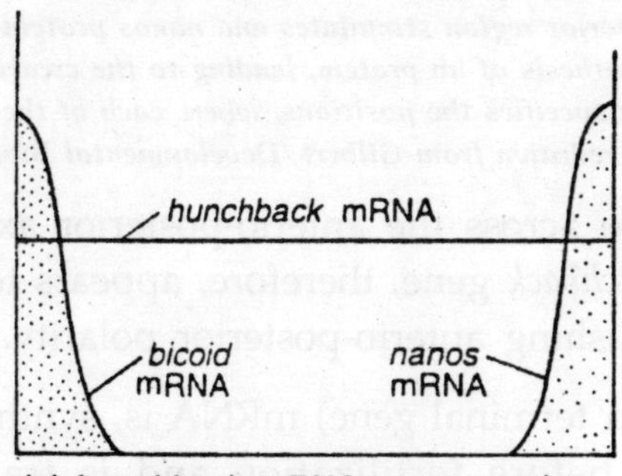

(a) hunchback mRNA concentration is same throughout the embryo before fertilization

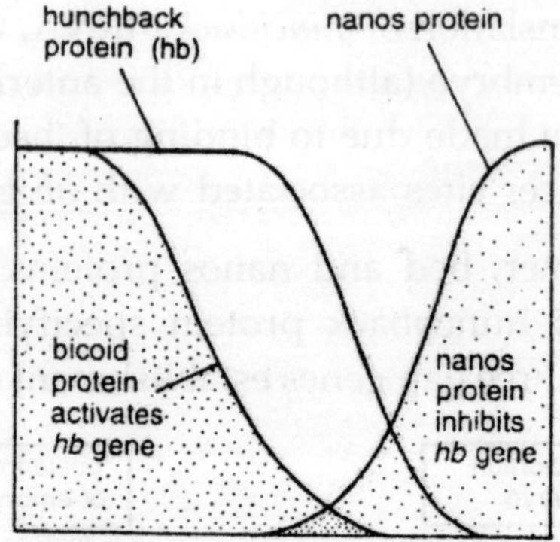

(b) gradient of hunchback protein after fertilization

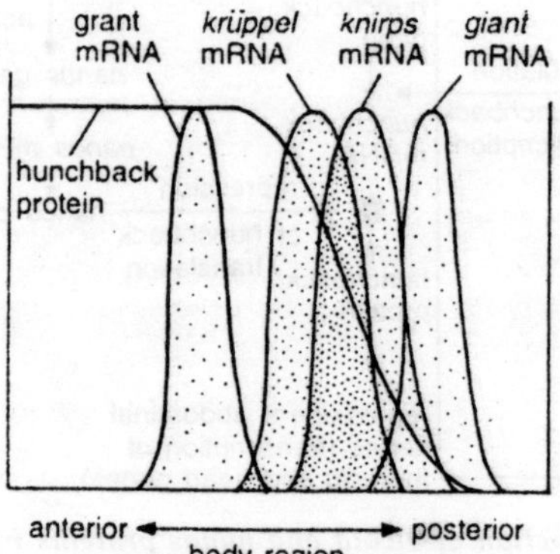

(c) concentration of hunchback protein specifying positions of transcription of three gap genes

Formation and function of hunchback (hb) protein gradient; note that bicoid protein in the anterior region stimulates and nanos protein in the posterior region inhibits synthesis of hb protein, leading to the creation of hb protein gradient, which specifies the positions, where each of the gap genes are transcribed (redrawn from Gilbert 'Developmental Biology', 1994).

hb protein also across the anterio-posterior axis of the egg (Fig.). The *hunchback* gene, therefore, appears to be the focal point in establishing anterio-posterior polarity.

The *torso* (a terminal gene) mRNA is, synthesized by the ovarian cells before fertilization and is translated after fertilization. The torso protein is localized in the cell membrane throughout the egg. The activator of *torso* protein is torsolike protein, so that a negative mutation in the *torsolike* gene gives phenotype, which is like *torso* mutant. Activation of *torso* initiates a pathway that eventually activates *tailless* and *huckebein* gap genes, responsible for activation of acron and telson specific genes. Terminal genes help formation of telson, but in the presence of bicoid (bcd) protein, they form an acron.

Genes Controlling Segmentation Pattern (Gap Genes, Pair Rule genes and Segment Polarity Genes, Expressed in the Zygote): In early embryo development, the gradients of morphogens play an important role and bring about cell *specification,* but not the irreversible *determination.* The transition from specification to determination is mediated by segmentation genes which are expressed after fertilization. These genes divide the early embryo into a series of repeating segmental primordia along the anterio-posterior axis. Mutations in these genes, therefore, lead to loss of certain segments or parts of these segments. There are three classes of these segmentation genes - *gap genes, pair rule genes* and *segment polarity genes* (Table). Mutations in these genes affect parasegments (Fig.). These parasegments are regions of embryo separated by mesodermal thickenings and ectodermal grooves, which divide the embryo into 14 regions. Parasegments do not become the segments of larva or the adult; instead each parasegment includes the posterior part of an anterior segment and the anterior portion of the segment behind it. The gap genes and pair rule genes are two classes of genes, which express zygotically before cellularization of blastoderm. The segment polarity genes, however, express after the cellularization starts.

Major Genes Affecting Segmentation Patterns in Drosophila

Gap genes	*Pair-rule genes*		*Segment polarity genes*
	Primary	*Secondary*	
Krüppel (Kr)	*orthodenticle (otd)*	*fushi tarazu (ftz)*	*engrailed (en)*
knirps (kni)	*hairy (h)*	*odd-paired (opa)*	*wingless (wg)*
hunchback (hb)	*even s'kipped (eve)*	*odd skipped (sdd)*	*cubitus interruptus (cb)*[D]

Contd...

Gap genes	*Pair-rule genes* *Primary*	*Secondary*	*Segment polarity genes*
giant (get)	*runt (run)*	*sloppy paired (sip)*	*hedgehog (hh)*
tailless (til)		*paired (prd)*	*fused (fu)*
huckebein (hkb)			*armadillo (arm)*
button head (btd)			*patched (ptc)*
empty spiracles (ems)			*gooseberry (gsb)*

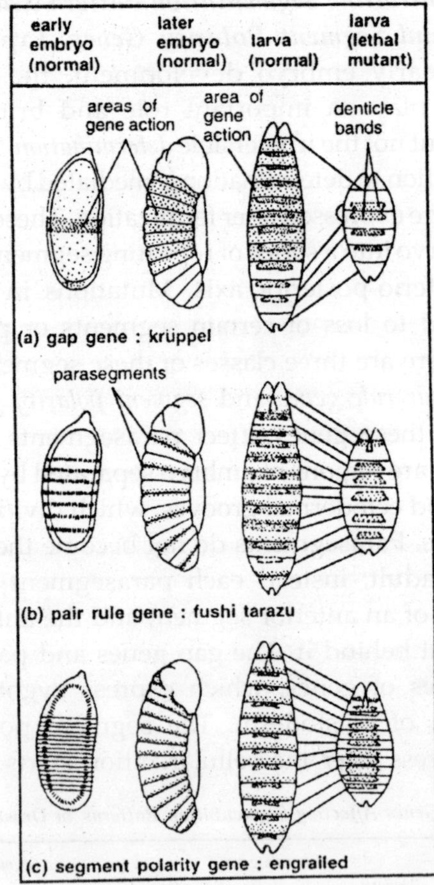

Three types of mutants (kruppel, fushi tarazu, engrailed) for segmentation pattern; in the three columns on the left, the shaded regions are shown, where the genes are expressed; in the fourth column on the right these regions are missing in the mutants.

Gap Genes : The gap genes (the first zygotically active segmentation genes) in their mutant form create gaps in anterio-posterior pattern and include *hunchback (hb), krupal (kr)* and *knirps (kni).* Their products have DNA-binding finger domains. The *bcd* and *osk* genes (particularly *nanos)* repress transcription of *kr* in the anterior and posterior regions respectively, thus restricting its expression to central region. Similarly *bcd* activates

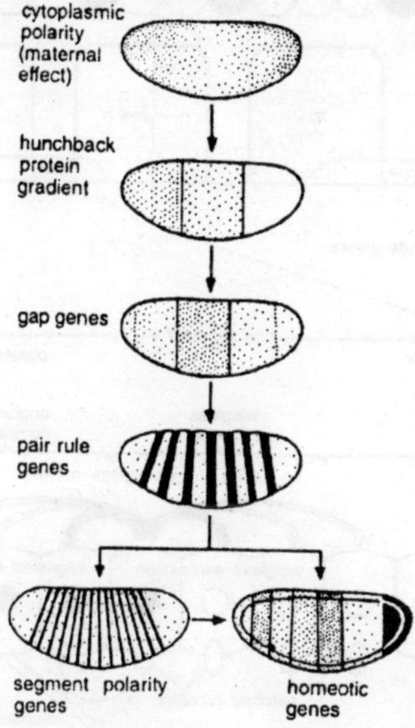

A model of striped pattern formation in Drosophila, showing stages of the expression of different classes of genes (redrawn from Gilbert 'Developmental Biology', 1994).

hb, and *osk* activates *kni.* Further, *kr* and *hb* repress each other. In the anterior region bicoid protein activates *hb* transcription; *hb* mRNA is translated into hb protein in the anterior region, since nanos protein, which inhibits translation of hb mRNA in the posterior region is absent in the anterior region. These conclusions are based on the study of a number of mutants. These gap genes, in turn, regulate the position specific expression of *pair rule genes;* the pair rule genes regulate the expression of segment polarity genes which in turn regulate the expression of *homeotic genes.*

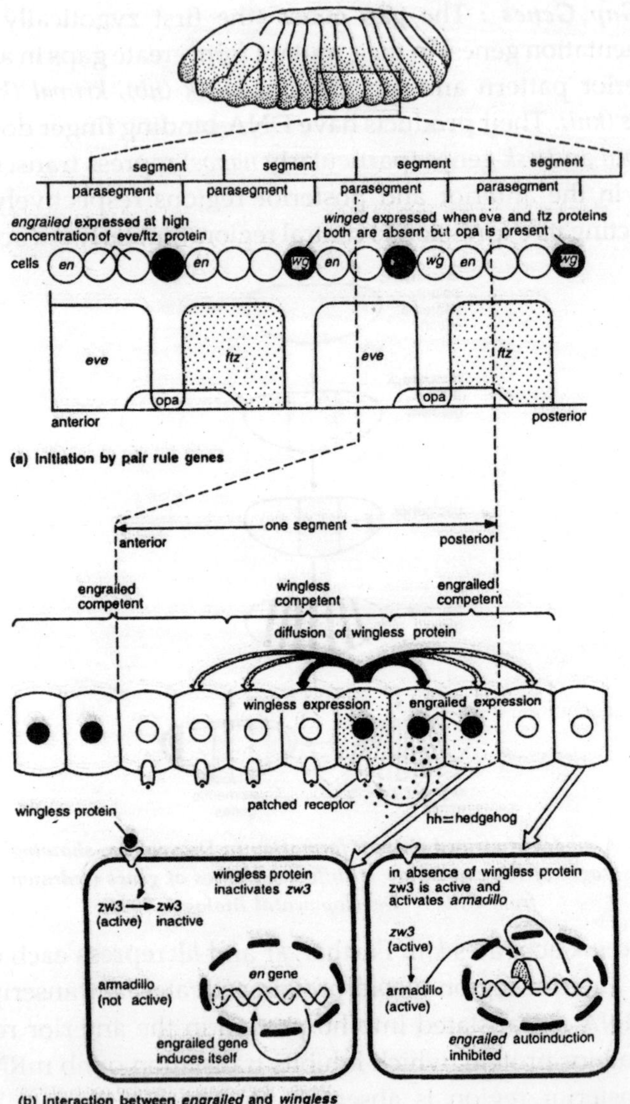

A Model for the transcription of segment polarity genes engrailed (en) and wingless (wg) (redrawn from Gilbert `Development Biology', 1994).

Pair Rule Genes : There are nine pair rule genes (which are expressed periodically), of which only two (*runt-run* and *hairy = h*) appear to have a major function in the generation of the striped patterns of their own expression and that of other genes of this class (hence called pair rule genes). Both genes, *run* and *h*, also regulate expression of each other. By the last interphase, before cellularization, transcripts of both genes (*run* and *h*) are already localized to two complementary series of seven bands that encircle the embryo. The generation of these striped patterns is a key event in the development (Fig.).

Segment Polarity Genes : The gap genes and the pair-rule genes discussed above are responsible for interactions within the syncytial embryo (*i.e.* up to 13th division cycle). But once the cells begin to be formed, interactions take place between cells. These interactions are controlled by certain genes called segment polarity genes, which establish the fate of individual cells within the segment. Two of the several segment polarity genes, that are important include the following: (i) *engrailed (en)* gene which is responsible for ensuring anterio-posterior boundary of each segment, and (ii) *wingless (wg)* gene, which is responsible for the proper expression of different regions of each segment.

The transcription of the segment polarity genes is regulated by pair rule genes. For instance, transcription of *wg* is repressed by products of *ftz (fushi tarazu)* and *eve (even skipped)*, and is stimulated by the product of *sdd* (odd skipped). Reverse is the situation with *en (engrailed)*, which is active in the presence of ftz protein or eve protein and is repressed by sdd (odd skipped) protein. Consequently, *wg* is transcribed in the cell directly anterior to the cells, where *en* is transcribed. The interactions between cells expressing *wg* and *en* are responsible for the maintenance of continued expression of *wg* and *en* (Fig.). The following interactions are involved:

(i) wingless protein diffuses to the adjoining cells and converts active inhibitory zw3 (zest-white 3) into inactive zw3, thus permitting autoinduction of *engrailed (en)*;

(ii) The cells synthesizing engrailed protein also synthesize hedgehog protein, which diffuses and binds to patched protein, thus inhibiting this protein, which otherwise inhibits synthesis of wg (wingless) protein.

Homeotic Genes : Once the embryo is divided into segments due to products of segment polarity genes, the products of gap genes, pair rule genes and segment polarity genes interact with another class of genes called homeotic genes. The transcription of these homeotic genes determines the developmental fate of each segment and mutation in any one of these homeotic genes leads to the transformation of one organ into another (*e.g.* antenna into leg). Most of the homeotic genes in *Drosophila* can be grouped into following two complexes:

(i) Antennapedia complex contains genes that affect head and thorax; it contains genes like *labial (lab), Antennapedia (Antp), Sex comb reduced (Scr), Deformed (Dfd)* and *proboscipedia (pb).*

(ii) Bithorax complex controls the posterior thoracic (T3) segment and the 8 abdominal segments (Fig.); it contains three genes: *ultrabithorax (ubx), abdominal A (abdA)* and *abdominal B (abdB); ubx* is required for the identity of the third thoracic segment, while *abdA* and *abdB* control the identity of 8 abdominal segments. Both these complexes are located on chromosome 3. An additional homeotic gene is *caudal,* which specifies telson.

The above homeotic genes were discovered due to bizarre phenotypes which were produced due to mutations in these genes. For instance, a dominant mutation in *Antp* (which normally expresses only in morax), leads to its expression in head as well as in thorax, so that the head segment behaves like a thoracic segment and legs would develop in place of antennae on the head (Figs.). Similarly a recessive mutation in *Antp* leads to failure of its expression in the second thoracic

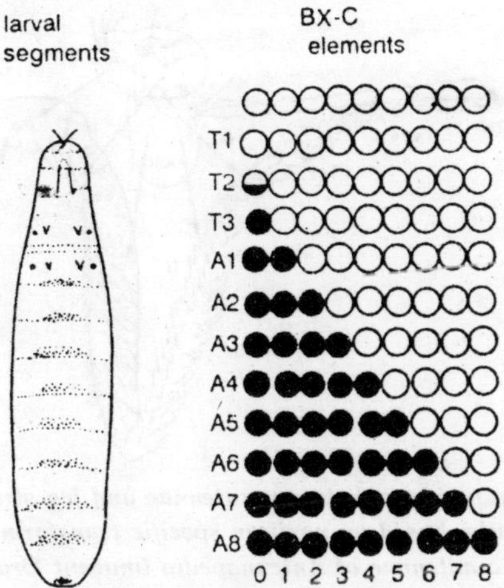

Model of function of bithorax complex; BX-C is divided into nine elements; the active elements are filled and inactive are open; the half filled circle represents Ubx* in the posterior compartment of T2 (note that in each progressively posterior segment, one more element is active).

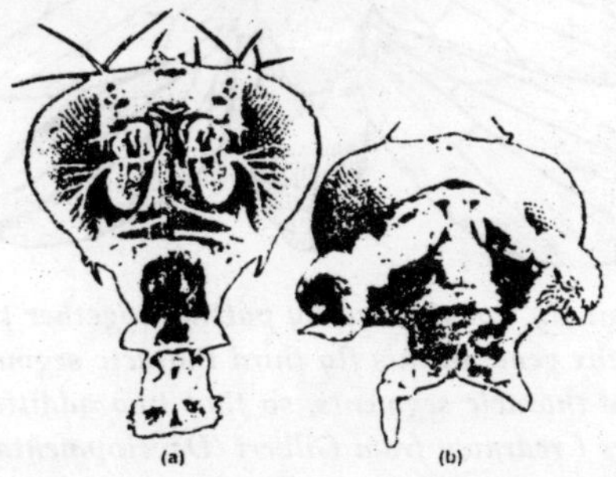

Head of a wild-type and an Antp mutant that converts antennae into egs (redrawn from Gilbert 'Developmental Biology', 1994).

segment, so that antennae will sprout in place of legs in this region. Similarly deletion of *ubx* (which is normally expressed in the 3rd thoracic segment) makes 3rd thoracic segment to behave like 2nd thoracic segment, so that it leads to development of wings in place of halters in the third thoracic segment (normally halters develop in the third and wings develop in the second segment), so that a fly with four wings instead of two wings (a characteristic of dipteran - *Drosophila)* will develop (Fig.).

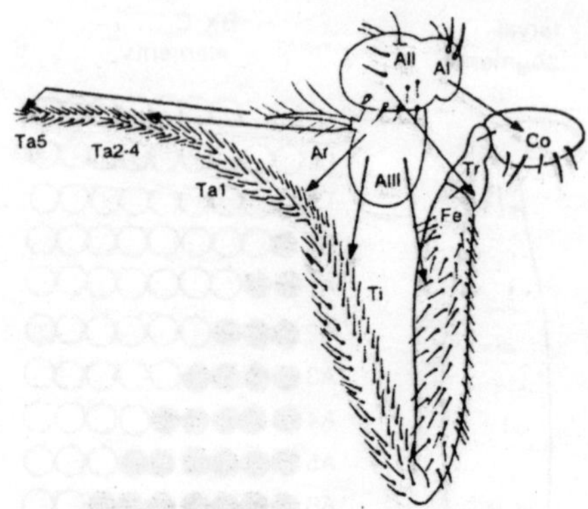

The correspondence between antennae and leg structures in Drosophila, based on position specific transformations in homeotic antennae of Antennapedia (mutant Drosophila).

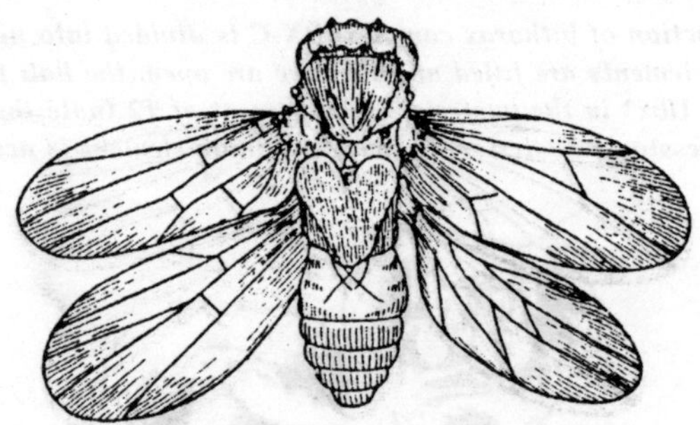

A four winged fruit fly, constructed by putting together three mutations in cis regulators of ubx gene; in this fly third thoracic segment is transformed into another second thoracic segments, so that two additional wings develop in place of halters (redrawn from Gilbert 'Developmental Biology', 1994).

Homeodomain Proteins : Homeodomain proteins are a family of transcription factors, characterized by a 60-amino acid domain that binds to certain regions of DNA. This homeodomain was first seen in proteins that are products of homeotic genes described above, but were later also discovered in other genes like *fushi tarazu*, *caudal* and *bicoid*.

The presence of some homeodomain containing proteins is also necessary for the determination of specific neurons. They are also necessary in determining the anterio-posterior axis of both invertebrates and vertebrates. Thus homeodomains are not restricted to proteins, which are the products of homeotic genes.

The homeodomain is encoded by the 180bp homeobox and specifies the binding sites for these proteins containing the homeodomain. This means that the binding property of these proteins depends on their homeodomain. The homeodomain folds into three α -helices, two of them folding into helix -turn-helix (HTH.) conformation characteristic of many transcription factors. The sites recognized by homeodomain are characterized by TAA(A)T motif.

It has been shown that the bicoid and Antennapedia homeodomains respectively use lysine or glutamine at position 9 to distinguish the corresponding binding sites. If lysine is replaced by glutamine in biocid, it starts recognizing Antennapedia binding site. The homeodomain proteins have multiple regulatory sites and the same gene can produce more than one homeodomain proteins due to mechanisms like alternate splicing and the same gene containing homeobox may be expressed in different cells due to multiple promoters.

Genes Controlling Dorso-ventral Polarity (Dorsal and Cactus Genes Maternally Expressed) : The dorsal-ventral axis of *Drosophila* egg is specified by atleast 20 genes, whose products are placed into the oocyte by the maternal genome. Absence of any one of eleven of these 20 genes causes loss of ventral structures. Absence of another gene, *cactus* causes even dorsal cell to become ventralised. Several steps are involved in the specification of dorsal-ventral axis and a critical step is the translocation of the 'dorsal' protein (a product of *dorsal* gene) from the cytoplasm to the nuclei of the ventral cells during the 14th division. This critical step requires a signal which is given by the ventral cells themselves.

After the translocation, the 'dorsal' protein specifies the different regions of the embryo. The dorsal protein activates *ventralising genes (snail* and *twisted)* and represses the dorsalising genes (*decapentaplegic* and *zerknullt*), so that the cells are developed into ventralised cells. If dorsal protein does not enter the nuclei of ventral cells, the ventralising genes will not be activated and dorsalizing genes will not be repressed, so that the cells will be dorsalised instead of being ventralised. Mutants lacking 'dorsal' protein, have all cells dorsalized and those having dorsal protein in every cell nucleus have all cells ventralised.

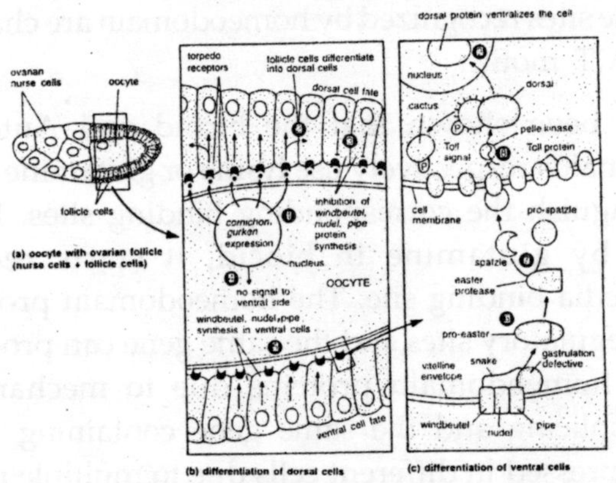

A model for the generation of dorsal-ventral polarity in Drosophila (for details see text); (redrawn from Gilbert 'Developmental Biology', 1994).

The signal for the translocation of 'dorsal' protein is mediated through an interaction between oocyte and the surrounding follicle cells (Fig.). The follicle cells appear to receive the dorsalizing signal through a receptor encoded by the *torpedo* gene. Maternal deficiency of this gene causes ventralization of the embryo. It is suspected that the *torpedo* encodes a receptor for which the ligand is encoded by two other genes, namely *gurken* and *cornichon.* Three other genes, namely *nude I (nd), pipe (pip)* and *windbeutel (wind)* are also needed in the follicle cells, but not in the oocyte. The absence of these genes in the mother causes the embryo to form only the dorsalized phenolype. These genes are turned off in the future dorsal cells by the activation of torpedo receptor. The complex formed by nudel, pipe and windbeutel proteins activates three serine proteases (snake, gastrulation defective and easter), which transduce the signal to the protein product of the *spatzle* gene. Spatzle protein is cleaved and the cleaved product binds to a receptor (product of *Toll* gene) in oocyte, thus transducing a signal (only on the ventral side, not on the dorsal side).

The signal from Toll receptor is translated by the products of the *tube (tub)* and *pelle (pll)* genes and helps in separation of 'dorsal' and 'cactus' proteins, which if remain-bound together do not allow the dorsal protein to be translocated from the cytoplasm to the nucleus. Dorsal and cactus proteins have been compared with NF-κB and IκB that are involved in differentiation of lymphocytes in mammals.

In the nucleus, the three target genes for dorsal proteins are *twist, snail* and *rhomboid* genes, expressed only in those nuclei of ventral cells, which receive the above 'dorsal' protein. The dorsal protein activates these genes, which in their turn activate or repress expression of mesodermal genes (activated by product of *twist)* and neuroectoderm gene like *rhomboid* (switched off by snail in mesoderm, but not in neuroectoderm).

Axes and Organ Primordia : The anterio-posterior and dorso-ventral axes of *Drosophila* embryos form a coordinate system that can be used to specify positions within the embryo, where specific sets of genes can be expressed for the formation of organ primordia. This can be illustrated by the formation of salivary gland in a region defined by (i) the vertical *Scr* expression band (no salivary glands are formed in *Scr* deficient mutants) and by (ii) horizontal region in the middle of the embryo's circumference, which lacks the products of *decapentaplegic* and *dorsal* genes.

Developmental Genetics of Higher Plants

Although several plant species have been used for a study of the genetics of development, the most advanced study pertains to the molecular and genetic analysis of flower development in *Arabidopsis thaliana* and *Antirrhinum majus.* The use of *Arabidopsis* as an experimental material has the following advantages :

(i) small genome size,

(ii) small duration of generation time,

(iii) ease of transformation and

(iv) small plant size. Similarly, the use of *Antirrhinim* has the following advantages :

   (i) well characterized transposable elements,

   (ii) large flowers,

   (iii) easy to emasculate and cross and

   (iv) ease of vegetative propagation.

The experimental approaches for the study of developmental biology in higher plants range over the fields of morphology, physiology, biochemistry and molecular biology. The genetic approach, coupled with the techniques of biochemistry and molecular biology offers unparalleled opportunities for the study of molecular basis of plant development. This genetic approach depends on the availability of suitable mutations.

During the last two decades a large number of mutations have been collected for the study of flower development (Table). Based on the nature of these mutations, they are classified in following classes :

(a) *Mutations that Affect Development of Flower Primordia (Class I Mutants) : Sterilis* and *steriloides* are two mutants, where abnormal inflorescences carrying only bracts, but no flowers are produced. However, occasionally few deformed flowers are produced in *steriloides* mutants. This suggests that these genes are involved in the development of flower primordia.

(b) *Mutations that alter the Symmetry of the Flower (Class II Mutants)* : Mutations that change zygomorphic flowers into actinomorphic or *vice versa* have also been identified. In *Antirrhinum, cycloidia* and *centroradialis* are two mutants, both conferring radial symmetry (actinomorphic flowers) on otherwise zygomorphic flowers. As reported above, *cantroradialis* gives radial symmetry due to the terminal position of the flowers, which are normally developed only in the axils of bracts.

(c) *Mutations that Alter Whorl Identity (Class III Mutants)* : A major class of whorl identity genes (which will affect the identity of organs in a whorl) includes those that affect the identities of organs in two adjacent whorls (Table).

In another classification, the genes controlling flower development can be broadly classified into following three groups : (i) floral induction genes and inflorescence architecture genes; (ii) floral meristem identity genes and (iii) floral organ identity genes. (Table).

Floral Induction and Inflorescence Meristem Identity Genes: The first step in flower development is floral induction, which involves a process where the vegetative phase changes to the reproductive phase. This is acçompanied by differentiation of vegetative shoot apex into an inflorescence axis bearing floral meristems, which develop into floral buds. The floral buds develop into flowers with whorls of sepals, petals, stamens and carpels. Once the carpels are formed, the floral shoot apex becomes non-meristematic and differentiates into tissues of carpel. Higher plants may differ among themselves in having determinate or indeterminate inflorescences. In plants with determinate inflorescences, the main axis terminates in a flower and additional flowers are produced below this flower on branches of the main axis. In plants with indeterminate inflorescences, on the other hand, the main axis is prevented

Genes Involved in the Regulation of Meristem and Floral Organ Identity in Arabiodopsis and Antirrhinum (first name is Arabidopsis Gene and Second name is Snapdragon Homologue)

Class of gene & name of gene	*Mutant phenotype*	*Molecular mechaniom*
I. Inflorescence Meristem Identity Gene		
TERMINAL FLOWER (TFL1); CENTRORADIAL1S (CEN)	Conversion of inflorescence to floral meristem; early flowering (only in *tfll)*	Negative regulator of *LFY, API* and *AP2*
II. Floral Meristem Identity Genes		
(i) *AGAMOUS (AG); PLENA (PLE)*	Indeterminate and repetitive pattern of floral organs- [(sepal, petal, petal)$_n$]	Transcription factor (negative regulator)
(ii) *LEAFY (LFY) ; FLORICAVLA (FLO)*	Partial conversion of floral meristem to inflorescence	Transcription factor; positive regulator of *AP3* and *PI*
(iii) *APETALA1 (API).; SQVAMOSA (SQUA)*	Axillary flowers	Transcription factor (positive regulator)
(iv) *APETALA2 (AP2)*	Similar to *apl* under short day	Negative regulator of *AG*, thus a positive regulator
(v) *CAULIFLOWER (CAL)*	*CAL* is wild type, but *apl cal* causes conversion of floral meristem to inflorescence	—

Contd...

Class of gene & name of gene	*Mutant phenotype*	*Molecular mechaniom*
III. Floral Organ Identity Genes		
(i) *AGAMOUS (AG) ; PLENA (PLE)*	Stamens and carpels replaced by petals and sepals respectively	Transcription factor ; negative regulator of *API, AP2* and *AP3*
(ii) *APETALAI (API) ; SQUAMOSA (SQUA)*	Sepals replaced by leaves; petals absent	Transcription factor
(iii) *APETALA (AP2)*	Sepals and petals replaced by leaves/carpels and stamens respectively	Negative regulator of AG
(iv) *APETALA (AP3); DEFICIENS (DEF)*	Petals and stamens replaced by sepals and carpels respectively	Transcription factor; positive regulator of *PI*
(v) PISTILLATA (PI); GLOBOSA (GLO)	Same as *ap3*	Transcription factor

from forming a terminal flower, so that it can grow indefinitely permitting more flowers to be produced on an elongated axis (*e.g. Antirrhinum*). Two classes of mutants showing deviation from wild type inflorescence have been detected in *Antirrhinum*:

(i) *floricaula (flo)* mutation (called *leafy* in *Arabidopsis)* leads to failure in transition from inflorescence meristem to floral meristem, so that proliferating inflorescence shoots develop in place of flowers. Two other mutations *squemata* and *squamosa* giving phenotypes similar to that of *floricaula (flo)* were reported. Mutations producing proliferating inflorescences have also been obtained in alfalfa, tomato, maize, etc.

(ii) *centroradialis* mutation has an effect opposite to that of *flo* and leads to the development of terminal flower converting indeterminate inflorescences into determinate type. These mutants are characterized as follows :

(a) the mutant produces a terminal flower, thus changing the indeterminate condition into determinate, which converts the normal plants into more bushy plants;

(b) about 10 axillary flowers are produced below the terminal flower; the terminal floral meristem is developmentally more advanced than the axillary floral meristems below it;

(c) the number of organs in each whorl of the flower and their arrangement (phyllotaxy) differ in terminal and axillary flowers;

(d) the terminal flowers exhibit radial symmetry while the axial flowers exhibit bilateral symmetry. A terminal flower *(tfl)* mutant in *Arabidopsis* can be compared with the *cen* mutant of *Antirrhinum* and has an additional feature of flowering earlier than the wild type plants.

The difference between terminal and axillary flowers suggests that in *cen* mutants, the apical meristem is differently programmed and that *cen* gene interacts with floral meristem identity genes. The morphological development of inflorescences in *cen* mutants of *Artirrhinum* has been studied through scanning electron microscopy (Breadley *et al.*, 1996).

The expression of *FLO* and *SQUA* genes in wild and *cen* mutants has also been studied. In wild type apices, within 1-2 days of floral induction, *FLO* is activated in the youngest primordia flanking the apical meristem. Thus, expression of *FLO* is an early marker for commitment to floral meristem identity. Its expression, therefore, could be used for comparing *FLO* induction in *cen* mutant with wild type, to find out the time of activation of the apex. Both in wild type and *cen* mutants, *FLO* was expressed in the primordia flanking the apical meristem within 1-2 days of floral induction (transfer to long days). Only after day 6, expression of *FLO* in the apical meristem was observed in *cen* mutants suggesting that apical meristem of *cen* mutants was recruited for floral development at a late stage, thus appearing more advanced than an axillary meristem. This also suggested that *CEN* prevents expression of *FLO* in apical meristem.

The *FLO* gene has been isolated and was shown to produce a transcript of 1.6kb, encoding a protein FLO of 396 amino acids showing some similarity with transcription factors. In situ hybridization shows that *FLO* is expressed from a very early stage in wild type inflorescence in a very specific temporal and spatial sequence. It expresses in bract primordia followed by expression in primordia for sepals, petals and carpels but not in the primordia for stamens. Therefore, FLO^{+} may be involved in transition from vegetative to floral axis and in the initiation of floral primordia. Absence of its expression, however, is required for normal stamen development. The FLO^{+} may also interact with other genes, like those meant for whorl identity.

In *Antirrhinum*, *CEN* and *FLO* expressions are actually coordinated. In *flo* mutants, flowers are replaced by indeterminate shoots, while in *cen* mutants, the shoot bears a terminal flower. It appears that *cen* and *flo* regulate each other.

Recently *CEN* gene was isolated by transposon tagging and a wild type 6.5kb *EcoRl* fragment was subcloned and fully sequenced. The gene consisted of four exons measuring 980bp. It was also shown that *CEN* is expressed in regions below the apical meristematic dome and does not overlap the region of *FLO* expression observed in *cen* mutants.

Floral Organ Identity Genes in Arabidopais Thaliana

Class of genes	*Gene & gene symbol (wild type)*	*Mutant allele and phenotype*
A	*APETALA1 (AP1)*	*ap1* (sepals replaced by bract like structures, petals aborted or sepaloid)
	APETALA2 (AP2)	*ap2* (sepals replaced by carpelloid structures, and petals replaced by stamens)
B	*APETALA3 (AP3)*	*ap3* (petals replaced by sepals and stamens replaced by carpels)
	P1STILLATA (PI)	*pi* (similar to *ap3*)
C	*AGAMOUS (AG)*	*ag* (stamens replaced by petals and carpels replaced by sepals)

Floral Meristem Identity Genes : On floral induction, the apical meristem changes from vegetative to reproductive phase producing a number of floral primordia. This is then called inflorescence meristem, which is indeterminate (unlimited growth) and gives a number of determinate (limited growth) floral meristems. Mutants have been obtained, where:

(i) the shoot fails to produce floral meristems, and those where

(ii) flowers are transformed into inflorescence shoots. Some of these mutants are listed in Tables.

Mutations in *LFY* and *API* result in the conversion of an individual floral meristem into an inflorescence meristem. Both of these genes are responsible for specifying floral meristem identity, but *LFY* plays a more predominant role. In *Antirrhinum* the gene corresponding to *LFY* is *FLORICAULA (FLO)* and mutations at this locus *(flo)* in *Antirrhinum* give a phenotype, which is more severely affected than in *lfy* mutants in *Arabidopsis. API* in *Arabidopsis* and *SQUA* in *Antirrhinum* are also required for specification of floral meristem identity, so that in mutants at these loci, flowers are transformed into branched structures bearing flowers. It has also been shown that *LFY* and *API* act synergistically.

The third gene *CAL* is only an enhancer of *API* alleles and mutations at this locus do not affect floral meristem identity. Another gene *APETALA2* or *AP2* also reinforces the activity of *LFY* and *API*. Thus, *LFY, API, AP2* and *CAL* promote formation of floral meristems. More recently, in rice, the gene *RFL*, a homologue of *FLO* of *Antirrhinum* and *LFY* of *Arabidopsis* has been isolated and studied (results published in Proc. Natl. Acad. Sci. USA, March 1998). It was shown that *RFL* differs from *FLO* and, *LFY*, and that its presence may not be essential for floral initiation in rice. Instead, it may be needed for panicle branching.. This suggests that the results obtained in *Arabidopsis* and *Antirrhinum* can not be generalized and applied to distantly related species such as rice and other grasses.

There are also genes, which determine the time of flowering and repress the formation of floral meristem. These include *TERMINAL FLOWER (TFL)* and *EMBRYONIC FLOWER (EMF)* genes of *Arabidopsis*. Mutations at the *TFL* locus result in earlier flowering and also in the transformation of the apical meristem into floral meristem. Similarly, mutation in *EMF* leads to appearance of a single flower soon after germination.

The *LFY, API, AP2* and *CAL* genes of *Arabidopsis* and *FLO* and *SQUA* of *Antirrhinum* have been cloned and sequenced.

By *in-situ* expression studies, it was shown that before the development of floral meristem, *LFY* transcript (RNA) is present on the flanks of apical inflorescence meristem but not on the apical meristem itself. *LFY* transcript is also seen in early floral primordia but is reduced in later stages. The *API* RNA is detected in very early floral primordia, slightly later than the onset of *LFY*. This early expression of *API* is later restricted to sepal and petal primordia, when the floral organ identity gene *AGAMOUS* is expressed. It is, therefore, believed that *API* is not only meant for floral meristem identity, but also for organ identity specifying sepal and petal development. *CAL* is expressed only early in floral primordia and therefore can substitute for *API* for meristem identity but not for organ identity. The expression pattern of *FLO* in *Antirrhinum* differs considerably from that of *LFY*, suggesting that *LFY* and *FLO* control flower development in different ways.

AP2 gene is expressed in the leaf primordia, apical meristem and in floral primordia. *AP2* is believed to be involved (as in case of *API)* not only in reinforcing meristem identity, but also in specification of sepal and petal primordia. *'Floral meristem identity genes'* like *LFY*, *AP2* and *CAL* actually work as a bridge between "floral induction genes" and "floral organ identity genes".

Floral Organ Identity Genes : After the inflorescence meristem and floral meristem are specified, the next step in floral development is the specification of the type, number and position of organs (sepals, petals, stamens and carpels). A large number of mutants for flower development having abnormal floral whorls have been isolated, which could be broadly classified into two groups :

(i) homeotic mutations that cause the transformation of one flower whorl into a whorl of a different type; several such mutants were examined in petunia, and are listed in Table and

(ii) meristic mutations that alter the number of a structure within a floral whorl (*e.g. clavata,* where number of carpels increased from two to three or four). A number of mutations including both these types, known in *Arabidopsis* are listed in Table.

Analogous mutations have been observed in other plant species, indicating that the fundamental processes controlling floral morphogenesis are highly conserved. In wild type plants, after a period of vegetative growth, the apex undergoes a transition to become an inflorescence meristem to produce indeterminate inflorescences, which do not produce a terminal flower. The inflorescence bears bracts, in the axils of which flowers develop.

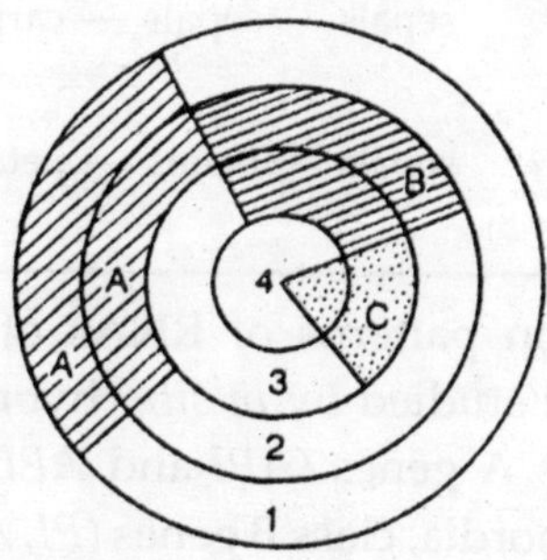

Three regions of floral meristem, i.e., A, B and C, such that genes (a, b and c) acting in these regions have regulatory functions; note that out of four whorls (1 to 4), whorl 1 is controlled by a alone; whorl 2 is controlled by a and b; whorl 3 is controlled by b and c and whorl 4 is controlled by c alone.

The floral mutants studied in *A. thaliana* are broadly classified into three categories (Table; Fig.), and were used to develop an '*ABC* model to explain flower development. According to this model, the homeolic genes can be broadly classified into three types A, B and C. Their function is explained by the following three assumptions or hypotheses:

(i) each gene functions in two adjacent whorls, so that A functions in 1st and 2nd whorls, B functions in 2nd and 3rd whorls and C functions in 3rd and 4th whorls;

(ii) 'A' functions to form sepals in the first whorl, A + B function to form petals in the second whorl, B + C function to form stamens in the third whorl and C alone functions to give carpels in the fourth whorl;

(iii) 'A' class and 'C class genes repress the function of each other. Using the above assumptions, phenotypes of the flowers in homeotic mutants can be explained (Table).

I'henotypes of Mutants based on ABC Model.

Mutation type	*Genotype of mutant*	*Phenotype (four whorls)*
a	*aBC* (apetala 2)	carpels — stamens — stamens — carpels
b	*AbC* (apetala 3)	sepals — sepals — carpels — carpels
c	*ABc* (agatnous)	sepals — petals — petals — sepals

When expression patterns of RNAs of the above organ identity genes were studied by *in-situ* 'hybridization, it could be shown that class A genes *(API* and *API)* are expressed in sepal and petal primordia, class B genes *(PI, AP3)* are expressed in petal and stamen primordia and class C gene *(AG)* is expressed in stamen and carpel primordia.

That A and C class geres function antagonistically to each other is supported by the observation that in plants mutant for class A gene *(ap2)*, expression of C class *(AG)* gene expands to whorls one and two. *API* plays a minor role of repressing *AG* expression in the 3rd and 4th whorls. In *ag* mutant flowers, *API* expression expands to the inner whorls of the flower. The early expression pattern of class B genes *(AP3, PI)* is independent of class A and C genes. The expression of *PI* requires *AP3* activity and *vice-versa*, implying a role of *AP3* in the maintenance of *PI* activity and that of *PI* in the maintenance of *AP3* activity.

Homeotic genes (*e.g. gp* and *bl)* have also been identified in other plant systems like *Petunia* but mutants for these *Petunia* genes are affected in one whorl only, while those in *Arabidopsis* and *Antirrhinum are* affected in two consecutive whorls. Some of these homeotic mutants identified in *Petunia* are listed in Table.

Homeotic Genes Identified in Petunia

Mutant	*Phenotype*
green petal (gp)	Petals replaced by sepals
blind (bl)	Petals replaced by stamens
double (Dot)	Increased number of petals and stamens in 2nd whorl
ph3	Carpels replaced by stamens (female sterile)

MADS - Box Gene Family in Higher Plants : The floral organ identity genes have been shown to encode transcription factors each with the conserved MADS-DNA binding domain (except *APETALA2).* The MADS domain has 56 amino acids and was initially found in four different proteins including transcription factor MCM of yeast, in proteins coded by *AGAMOUS* and *DEFICIENS* genes of *Arabidopsis* and in the transcription factor SRF of mammals, hence the name MADS to represent the first four members of this class of transcription factors. The MADS domain is known to work as DNA- binding domain and also helps in protein -protein interactions. The homeotic MADS box genes are actually a subset of a larger MADS box gene family, some of whose members are expressed in all tissues, whereas others display a tissue specific or organ specific expression.

A number of MADS box genes have also been identified in *Petunia.* They are listed in Table, showing their whorl specific expression.

Expressoen Patterns of MADS box Genes in Petunia (for each gene, first row shows expression in wild type and second row shows expression in gp mutants; - no detectable expression, + low expression, ++ intermediate expression, +++ high expression)

Gene product	Expression Whorl 1	Whorl 2	Whorl 3	Whorl 4
pMADS 1	+	+++	+++	–
	–	–	–	–
pMADS 2	–	+++	++	–
	–	–	++++	–
FBP 1	–	+++	++	–
	–	–		–
pMADS3	–	–	++	+++
	–	–	++	+++
pMADS4	+++	+++	+	+++
	+++	+++	+	+++

Gene SUPERMAN (SUP) in Arabidopsis : In *Arabidopsis,* the gene *SUPERMAN (SUP)* is necessary for the proper spatial development of reproductive floral organs. Recessive mutations *(sup)* cause extra stamens to be formed as the fourth whorl in place of carpels. For the expression of *sup* phenotype, ectopic expression of wild type *AP3* and *PI* (class B of function genes) is necessary, so that the double mutants *(ap3 sup* and *pi sup)* exhibit the phenotypes of single mutants *(ap3, pi).* This had suggested that *SUP* is an upstream negative regulatory sequence, which will suppress the expression of *AP3* and *PI.* More recently, it was shown that *SUP* actually expresses in the third whorl. The SUPERMAN protein contains one Zn-finger and a region resembling Ieucine zipper suggesting that it may function as a transcription factor.

Transgenic Plants for Study of Floral Development : Transgenic plants of *Arabidopsis* have also been produced for the study of the function of genes, belonging to three classes (A, B, C) as proposed in ABC model. For instance in an *ag* mutant (mutant for class C gene), when class B genes *(PI* and *APS)* were introduced, all the four whorls had petals.

Transgenic plants have also been produced by introducing into wild type plants *LEAFY (LFY)* and *APETALA1 (API)* genes carrying promoters permitting constitutive expression. Both transgenes induce conversion of shoot meristem into floral meristem, so that the transgenic plants are early flowering. Similarly, when *LFY* gene was introduced into aspen plants, it flowered within six months, though generally it takes 6-7 years for flowering. These observations confirm that either of the two genes *(LFY* or *API)* is sufficient for the formation of floral meristem.

5

Genetic Recombination

The term recombination has often been treated as a synonym to crossing over. However, it should be recognized that recombination covers a much wider spectrum of events including, (i) crossing over, both meiotic and mitotic (reciprocal exchange of segments), (ii) gene conversion (non- reciprocal events), (iii) exchange between sister chromatids (even though this does not lead to a change in genotype), (iv) repair of DNA damage, etc. For the purpose of this chapter, in the treatment of recombination, we have excluded processes, in which different DNA sequences are simply legated together to produce, what is commonly called recombination DNA. The recombinant DNA and gene cloning will be separately.

The mechanism of genetic recombination is mainly based on breakage and reunion process, which has been demonstrated experimentally both at the level of chromosomes as well as at the level of DNA molecules. However, this simple breakage and reunion of chromatids or DNA molecules can not explain gene conversion and associated phenomena. Gene conversion leads to the production of rare octads with ratios like 6 : 2,

2 : 6, 5 : 3, etc. among 8 ascospores of a single ascus in ascomycetes like *Neumspora.* These results could not be explained even on the basis of copy choice model used earlier for explaining recombination events in bacteria and viruses, because this model does not take into account the semi-conservative method of DNA replication. Therefore, formation of hybrid DNA or heteroduplex DNA (DNA having two strands from different sources) has been assumed in several recent models proposed to explain the mechanism of genetic recombination at the molecular level. These models will be discussed in a relatively greater detail in this chapter.

Hybrid DNA Models Involving Single Strand Breaks (e.g. Holliday Model) : Several hybrid DNA or heteroduplex DNA models to explain the mechanism of recombination have been proposed which assume breaks in only one of the two strands of each of the two DNA molecules, belonging to two non-sister chromatids. Two such models were suggested, one each by H.L.K. Whitehouse and Robin Holliday, the major difference in these two models being in the polarity of the two strands undergoing single breaks.

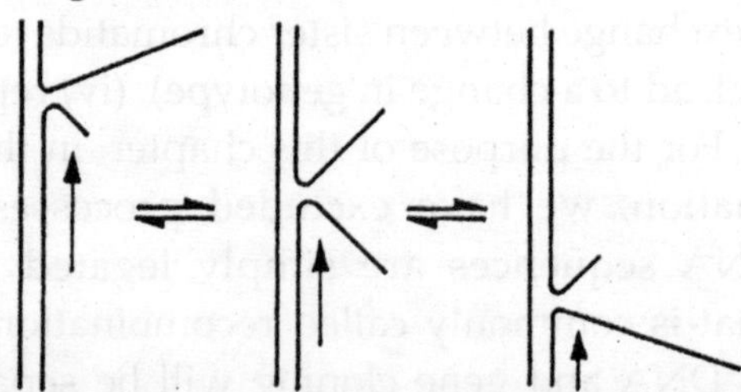

The phenomena of branch migration, which can occur in either direction, when an unpaired single strand displaces a paired strand.

In the model proposed by Whitehouse, these two strands will have opposite polarity (5' → 3' and 3' → 5'). but in Holliday's model, their polarity will be similar. Since Holliday's model received wide support and has been variously modified and discussed, this will be briefly described here. The model has been widely accepted both for eukaryotes and prokaryotes.

A paired bivalent in meiosis consists of four chromatids and at a particular location, two of them take part in recombination, although for double and multiple crossovers, three or all the four chromatids may be involved. The process starts with breakage, at the corresponding points, of the homologous strands of two paired DNA duplexes, each representing a chromatid, the other two chromatids being not involved. The breakage allows movement of the free ends

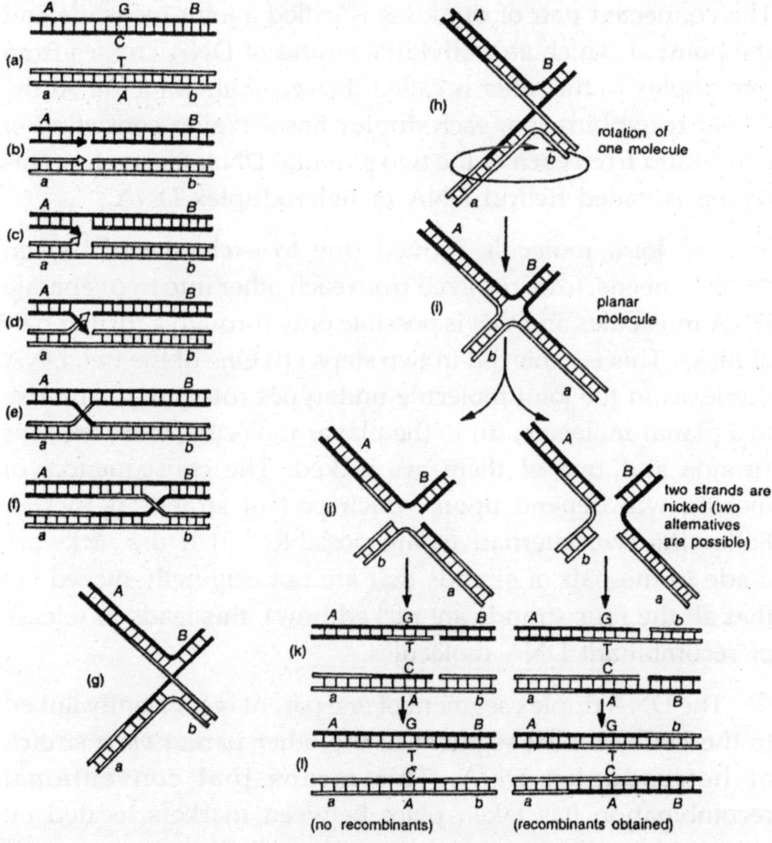

A hybrid DNA model for the molecular mechanism of recombination; the recombination between two paired duplex DNAs could be initiated by reciprocal single strand exchange (after single strand breaks), which is extended by branch migration and resolved by nicking.

created by the nicks. Each strand at the broken end leaves its partner, crosses over displacing the broken strand of the other non-sister molecule and pairs with its complement in the other DNA duplex (by a phenomenon called branch migration; see Fig.). It creates a connection between the two DNA duplexes. Initially this connection is sustained by hydrogen bonding, but at some point during branch migration, the nicks at the sites of exchange are sealed with the help of ligase enzyme.

The connected pair of duplexes is called a joint molecule and the point at which an individual strand of DNA crosses from one duplex to the other is called the recombinant joint. At the site of recombination, each duplex has a region consisting of one strand from each of the two parental DNA molecules. This region is called hybrid DNA or heteroduplex DNA.

The joint molecule formed due to exchange of broken strands, needs, to be resolved from each other into two separate DNA molecules and this is possible only through a further pair of nicks. This is achieved in two steps : (i) One of the two DNA duplexes in the joint molecule undergoes rotation, giving rise to a planar molecule, (ii) In the planar molecule, there are four strands and two of them are nicked. The consequences of nicking will depend upon, which pair of strands is nicked. Following two alternatives are possible : (i) If the nicks are made in the pair of strands that are not originally nicked (so that all the four strands are nicked now), this leads to release of recombinant DNA molecules.

The DNA duplex segment of one parent is covalently linked to the DNA duplex segment of the other parent *via* a stretch of heteroduplex DNA. This means that conventional recombination has taken place between markers located on either side of the heteroduplex region, (ii) If the two strands involved in the original nicking are nicked again, then only original parental duplexes with heteroduplex segments, but with no recombinant DNA molecules, are released. Different steps envisaged in Holliday's hybrid DNA model of recombination are shown in Figure.

Hybrid DNA Models Involving Double Strand Breaks : In this model, a double strand break instead of a single strand break is assumed. The double strand break caused by an endonuclease is enlarged, perhaps due to exonuclease activity, which nibbles away one strand on either side of the break

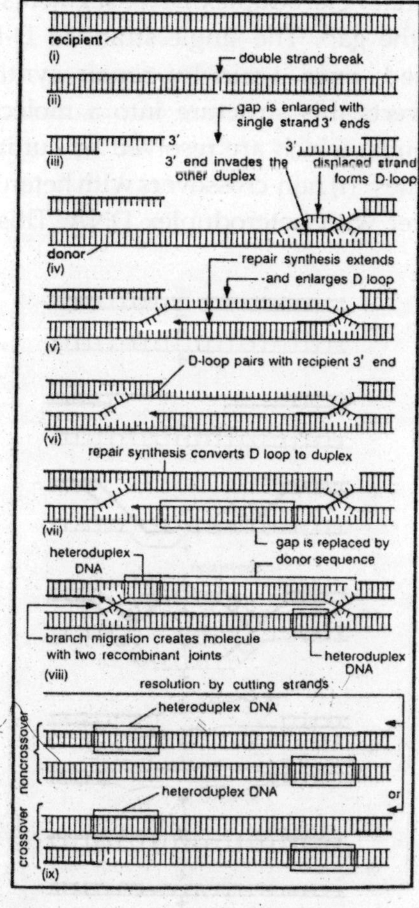

A hybrid DBA model for recombination, involving double strand break m, followed by formation of single stranded 3′ ends, only one of which mirates to a homoduplex DNA.

generating 3' single stranded termini. One of the two broken single strands invades the other duplex, so that he duplex molecules become connected by a stretch of heteroduplex DNA (models are also available where both broken single strands of a duplex invade the other duplex; see Fig.). The formation of heteroduplex DNA generates a D-loop, which is extended by repair synthesis, to cover the entire length of the gap. The extended single strand anneals with complementary sequence at the end, so that heteroduplex DNA segments are found on either side of the gap. The single stranded D-loop becomes double stranded, once again by repair synthesis. Branch migration converts this structure into a molecule with two recombinant joints, which are resolved by cutting. This leads to two possibilities : (i) non-crossovers with heteroduplex DNA; (ii) crossing over with heteroduplex DNA. This is illustrated in Figure.

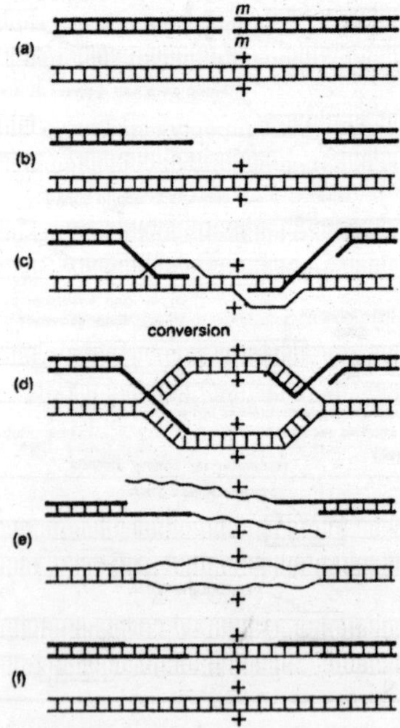

A hybrid DNA model for repair of double stranded discontinuities in DNA, leading to gene conversion.

The major difference between the models involving single and double strand breaks is that in the model with double strand breaks, the region between heteroduplex segments has the sequence of donor DNA in both molecules (this is why initiating DNA molecule is called recipient). Further, in double strand break model, there is initial loss of information, which is retrieved by resynthesis of DNA. In single strand break model, no such loss of information is involved. This double strand breaks model is consistent with events in yeast, where initiation of recombination is associated with receiving genetic information and where double-strand breaks are known to be involved in recombination like events.

Mismatch Repair and Post-meiotic Segregation (pms) in Heteroduplex Region : In the two models for recombination discussed above, it has been shown that heteroduplex DNA regions are produced, which are associated with flanking markers that may or may not have undergone crossing over. The formation of heteroduplex segments is the only plausible means of interaction between DNA duplexes. It also provides an explanation for the phenomenon of gene conversion at a locus, which thus, may or may not be associated with crossing over between the markers flanking this locus undergoing gene conversion. If the region of heteroduplex formation is homozygous at the locus under study, this will not be subjected to any gene conversion, because the strand of a duplex, although derived from non-sister chromatids, are still similar (no mismatch). But if there is heterozygosity at the locus under study and it falls in the heteroduplex or hybrid DNA region, then there will be mismatch between two strands of a DNA molecule. This may then lead to gene conversion in either of the two directions (+/a → +/+ or *a/a*). This mismatch actually means non-complementary base pairs, so that the heteroduplex DNA segment will be unstable. This mismatch will be recognized by the mismatch repair system in the same manner as thymine dimmers, induced by UV light, are recognized by an excision repair system.

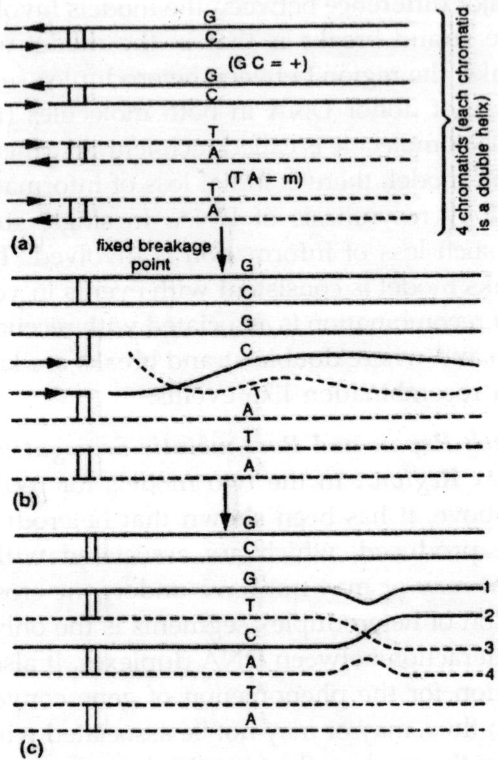

Origin of mismatches in heteroduplex DNA during recombination; (a) four chromatids of two paired homologous chromosomes carrying different alleles (+ = GC; m = AT); (b) breakage of single strands, which cross over to join at one another's breakage point; (c) formation of half chromatid chiasma and two stretches of neteroduplex DNA, (carrying mismatches-GT and CA) extending between breakpoints and chiasma.

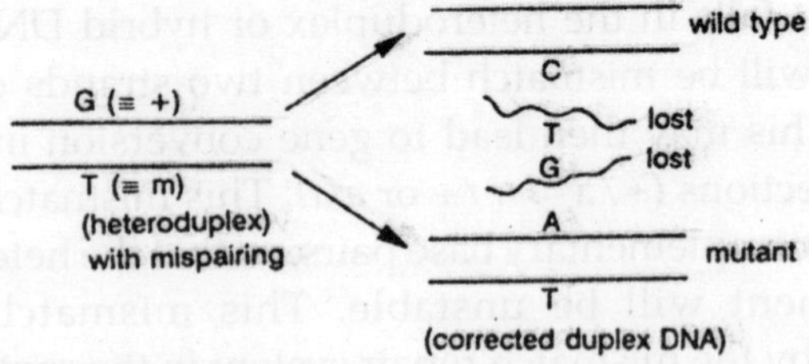

Correction of mismatched nucleotides in heteroduplex DNA to either wild type or mutant type, the mismatched nucleotides produce a distortion, that will be recognized by a repair system.

The above may be illustrated using a hypothetical example, where a heterozygote is *+lm* in which + and *m* differ by single base pair, their being G—C base pair in + allele and A—T base pair in *m* allele at the corresponding position. At meiosis, each of the four chromatids belonging to two paired homologous chromosomes, will have one DNA duplex thus making four DNA duplexes, of which, at a specific point, only two are involved in recombination process.

After interaction of DNA molecules, in heteroduplex segments, mismatches may by represented by G—T pair in one duplex and by C—A base pair in the other (Fig.). If we consider the fate of only one base pair *i.e.* G—T, this mismatch may be repaired to restore G—C base pair *i.e.* + allele or may give rise to A—T base pair *i.e. m* allele (Fig.).

If no mismatch repair is affected, then the mismatch may be corrected through replication at the first post meiotic mitotic division giving post meiotic segregation. Therefore, the heteroduplex segments may have any one of the following fates shown in Figure Table :

(i) no mismatch repair or correction, leading to aberrant 4 : 4 segregation with two mixed spore pairs *(mm, m+, m+, ++)*; this represents post-meiotic segregation;

(ii) mismatch repair or correction in one heteroduplex DNA giving 3 : 5 ratio *(mm, mm, m+, ++)*, or 5 : 3 ratio *(mm, m+, ++, ++)*; this represents, both gene conversion and post meiotic segregation (pms);

(iii) mismatch repair in both heteroduplexes giving a 2 : 6 ratio *(mm, mm, mm, ++)* or 6 : 2 ratio *(mm, ++, ++, ++)*; this represents gene conversion without pms. Other uncommon ratios are also found and can be explained by mismatch repair at one or both heteroduplex segments in either of the two directions.

Different ratio in the octad, obtained due to lack of correction or correction in one or both heteroduplex DNA molecules.

DNA strands before meiosis		DNA strands at hetero-duplex stage	No correction	Correction in one hetero-duplex (m→+)	Correction in both hetero-duplex DNAs (m→+)
m		m	3m	3m	2m
m		m			
m	hetero duplex	m			6+
m		+	1+	5+	
+	hetero duplex	m	1m		
+		+	3+		
+		+			
+		+			
Ratios			aberrant 4 : 4	5 : 3	6 : 2

Thus the phenomenon of gene conversion provided actual clues for the formulation of hybrid or heteroduplex DNA model for explaining the mechanism of genetic recombination or crossing over at the molecular level. Several features or events involved in the hybrid DNA models have been experimentally demonstrated *in vivo* or *in vitro*. These include:

(i) nicking,

(ii) strand displacement,

(iii) branch migration,

(iv) repair synthesis and

(v) legation.

Recombination intermediate structures envisaged in Holliday's model and other hybrid DNA models have also been observed under the electron microscope in recombining phages and plasmids. Figure. shows how two duplex circles

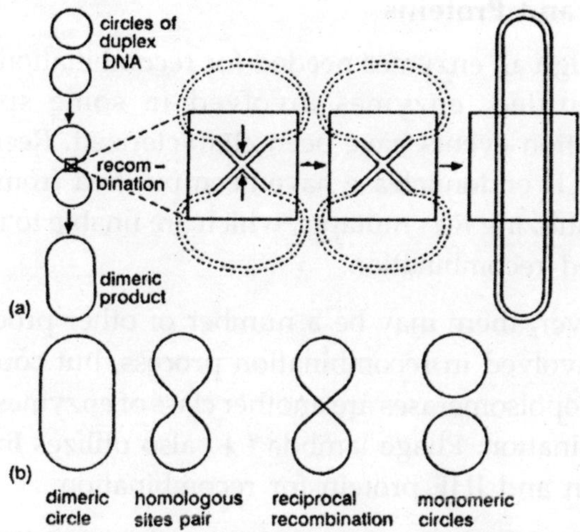

(a) Formation of a dimeric circle from two duplex circles through recombination, involving an intermediate structure resembling a figure of eight (8); (b) formation of two monomeric circles from a dimeric circle by recombination between homologous sequences.

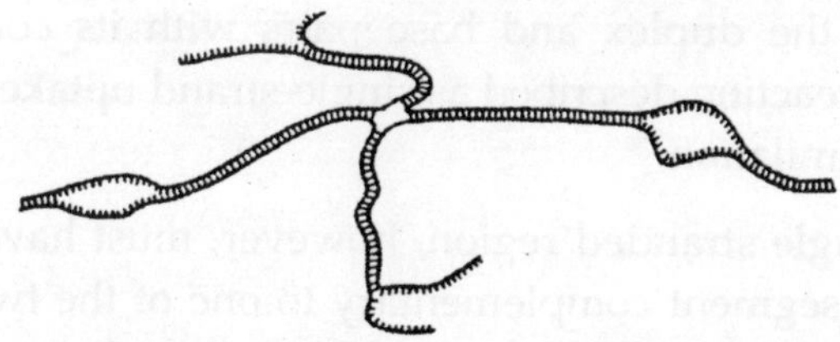

A chi structure with four duplex arms held together at the site of strand exchange (based on an electron micrograph).

(as found in plasmids) may give rise to a dimeric circle through an intermediate structure resembling a figure of eight (8) and how a dimeric circle can give rise to two monomeric recombinant circles. Structures like figure of eight may be isolated and will give rise to chi (χ) structures due to cleavage with a restriction enzyme, that cuts each monomeric circle once. This structure will have four arms held together by a region of heteroduplex DNA at the point of fusion of four arms. Such a chi (χ) structure has actually been observed see Figure.

Enzymes and Proteins

Although all enzymes needed for recombination have not been identified, enzymes involved in some specialized recombination events have been characterized. RecA protein and RecBCD endonuclease have been purified from bacterial systems, utilizing *Rec*⁻ mutants, which are unable to undertake generalized recombination.

However, there may be a number of other products that may be involved in recombination process, but could not be isolated. Topoisomerases are another class of enzymes involved in recombination. Phage lambda (λ) also utilizes Int protein, Xis protein and IHF protein for recombination.

RecA and RecBCD proteins promote base pairing between a single strand of DNA and its complement available in double stranded DNA molecule. The single strand displaces its homologue (strand which is similar to the invading single strand) in the duplex and base pairs with its complement through a reaction described as single-strand uptake or single-strand assimilation.

The single stranded region, however, must have a free 3' end and a segment complementary to one of the two strands of the duplex. Some forms of this reaction are shown in Figure. It is believed that the RecA protein first binds to single stranded DNA and then binds double stranded DNA and looks for the region complementary to single stranded DNA. This single-strand assimilation seems to be related to the initiation of recombination.

The models for recombination earlier described in this chapter, involve an intermediate stage, where both single strands cross over from one duplex to the other. RecA protein may presumably catalyse this reaction, even though such a protein has not so far been isolated from a eukaryotic system.

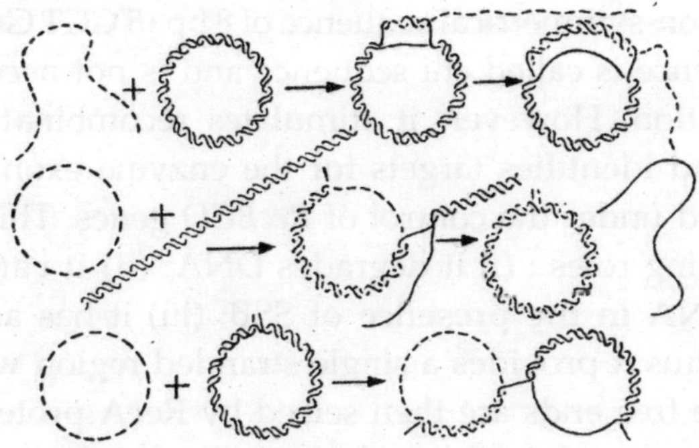

Different forms of reaction for invasion of single strand into a duplex, with the help of Rec A.

There are some other attributes of RecA, which make its role in recombination understandable. These include the following : (i) ability to polymerize single strands; (ii) ability to incorporate duplex DNA into filaments, giving duplexes longer than the original duplex. In view of this the reaction involving generation of heteroduplexes during recombination may involve following three steps : (i) a slow presynaptic phase, in which RecA polymerizes on single stranded DNA; (ii) a fast pairing reaction between single stranded DNA and its complement in double stranded DNA, to produce heteroduplex DNA; (iii) a slow displacement of one strand from the duplex, to produce a long region of heteroduplex DNA. The initial region of heteroduplex DNA may not be found in a double helical form (plectonemic), but the two strands may be associated side by side (paranerhic joint). This paranemic joint is unstable and further progress in recombination events requires its conversion to double helical form.

When the reacting DNA molecules are circular, the product of RecA sponsored reaction is a chi (χ) structure. Although action of RecA is independent of DNA sequence, certain hot spots stimulate the RecA recombination system. These hot spots were discovered in lambda phage *chi* mutants and share common non-symmetrical sequence of 8 bp (5'GCT GG TGG3')- This sequence is called chi sequence and is not necessary for recombination. However, it stimulates recombination in its vicinity and identifies targets for the enzyme exonuclease V synthesized under the control of *RecBCD* genes. This enzyme has following roles : (i) it degrades DNA; (ii) it can unwind duplex DNA in the presence of SSB; (iii) it has an ATPase activity. Thus it provides a single stranded region with a free end. These free ends are then seized by RecA protein, which requires free termini to start the strand assimilation reaction.

Topoisomerases : Recombination, like replication, also requires topological manipulation of DNA, which is facilitated by topoisomerases. These enzymes may relax or introduce supercoils in DNA and are required to disentangle DNA molecules that have become catenated by recombination or replication. Topoisomerases can be of Type I (making transient break in one strand) or Type II (making transient double strand breaks). Type I topoisomerases (*e.g.* product of *topA* gene of *E. coli)* relax negatively supercoiled DNA, while Type II topoisomerases (*e.g.* gyrase) relax both negative and positive supercoils. The removal of supercoils from the crossing segments is required during the recombination process.

6

Chemistry of the Gene

Nucleic acids as one of the several classes of macromolecules are found in a cell. Of the two main types of nucleic acids (DNA and RNA), DNA is known to be the genetic material in most eukaryotes. DNA, therefore, is the most important of all the macromolecules of the cell. All cells of a living organism possess the same quantity and quality of DNA.

Since these cells result from a single cell by successive cell divisions, a very precise mechanism for DNA synthesis (replication) should be available in the cell. This DNA is also variously modified to allow for the recognition of (i) old and newly synthesized strands of DNA duplex, (ii) any foreign DNA invading the cell or (iii) the active (expressed) and passive genes. There are also mechanisms for repair of DNA damage in the cell. In this chapter, details of the mechanism of DNA synthesis (particularly DNA directed DNA synthesis *i.e.* DNA replication, and briefly also RNA directed DNA synthesis *i.e.* reverse transcription) along with a very brief account of DNA modification (methylation, etc.) and DNA repair will be presented.

DNA Replication : General Features

Semi-conservative DNA Replication in E. coli : Both DNA as well as genetic RNA are capable of undergoing self-replication. Theoretically, replication of double stranded DNA, could be (i) conservative, (ii) dispersive, or (iii) semi-conservative. The conservative replication would mean that double stranded molecule is conserved as such and that a new copy is synthesized from old molecule. In dispersive replication, the old molecule should disintegrate and two new molecules would be synthesized. In the semi-conservative replication, the two strands would separate from one another, maintain their integrity and each will synthesize, from the pool of nucleotides, its complementary strand. The result would be, that the newly synthesized molecule would carry or conserve one of the two strands from the parent molecule and the other strand would be newly assembled (Fig.). There is sufficient evidence to prove that double stranded DNA really replicates by semi-conservative method. Meselson and Stahl's experiment (using N^{15}). M. Meselson and F.W. Stahl in 1958 reported the results of an experiment which was designed to test whether double stranded DNA replicates in a semi-conservative manner.

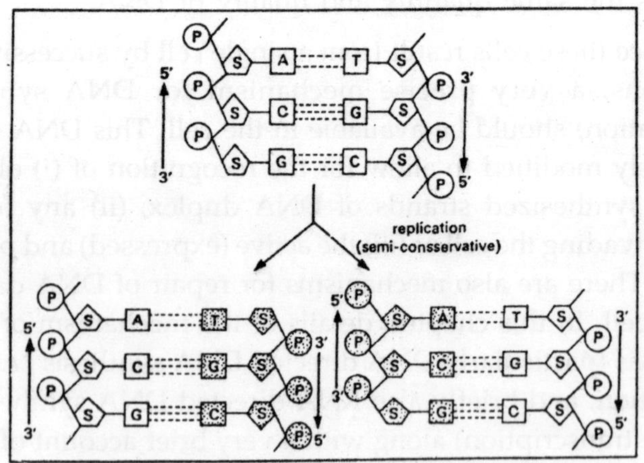

Basic plan of the structure of double helical DNA and the semi-conservative replication.

The following two basic principles were involved : (i) DNA was labelled with heavy nitrogen (N^{15}) and was then allowed to replicate in a medium containing N^{14}. If replication is semi-conservative, then after first generation of replication, one of the two strands would have heavy nitrogen (N^{15}) and the other strand would have normal nitrogen (N^{14}). The resulting molecule would have a density which is intermediate between N^{15} DNA and N^{14} DNA. This density will gradually fall and in the following generations of replication, density would approach that of N^{14} DNA. (ii) Another principle utilized was the preparation of a caesium chloride density gradient. This density gradient is prepared by gradual dilution of a heavy salt solution. When this is subjected to ultracentrifugation, with a substance having density within the range of this gradient, this substance will find place at its own level of density. This enables detection of very slight differences in density.

Meselson and Stahl (1958) allowed *Escherichia coli* cells to grow on N culture medium for about 14 cell generations, so that almost all nitrogen (N^{14}) in DNA is replaced by N^{15}. Then the cells were abruptly transferred to N^{14} culture medium. Since the time required for one cell generation was determined to be about 30 minutes, it was possible to remove the cells after one or more known number of generations of replication. Their DNA could then be analysed and the following results were obtained.

DNA sample obtained one cell generation after the transfer to N^{14} culture medium showed only one density band as observed through ultraviolet absorption pattern. This band indicated uniform homogeneous density of DNA after one cell generation. Moreover, the band was exactly between the bands formed by N^{15} DNA and N^{14} DNA, indicating that all DNA found after first generation had intermediate density. This is what one would expect if DNA replicates in a semi-conservative manner.

Density of DNA was studied in the following generations also. After two generations, when DNA was analysed, two bands were observed, as one would expect according to semi-conservative method of replication. These two bands were of equal intensity after the second generation. In the subsequent generations, although the same two bands appeared, the intensity of hybrid density band gradually decreased, while the intensity of light density band gradually increased (Fig.). Muselson and Stahl's experiment is, thus, a classical experiment which was the first to show that DNA replicates in a semi-conservative manner.

Cairns' Autoradiography Experiment : Semi-conservative mode of replication of bacterial chromosome was also demonstrated by J. Cairns, using the technique of autoradiography. In autoradiography technique, the cells are first supplied with a suitable radioactive material like tritiated thymidine (H^3-TdR; H^3 is heavy isotope of hydrogen and it replaces normal hydrogen in thymidine to give rise to tritiated thymidine).

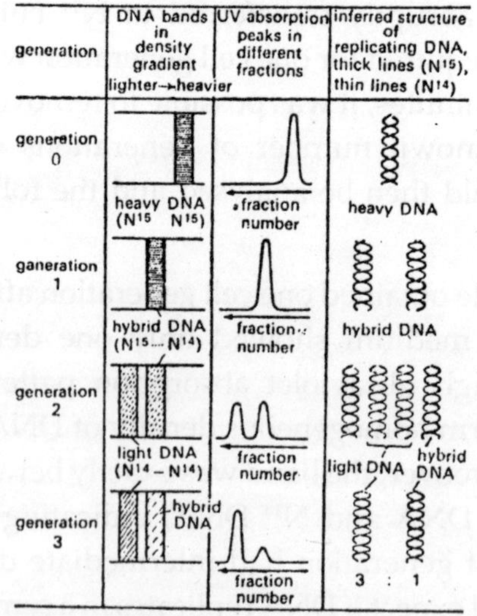

Experiment of Meselson and Stahl demonstrating semi-conservative replication.

Tritiated thymidine is used since this will selectively label only DNA and will not label RNA, since thyinine base is absent in RNA. The tritiated thymidine gets incorporated into DNA and replaces ordinary thymidine. The cellular material is then sectioned or else the cells may be broken down to release the intact bacterial chromosomes on slides. These slides are then covered by photographic emulsion and stored in dark. During this storage, the particles emitted by tritiated thymidine will expose the film, which can be developed. This photograph will then show the regions of the presence of tritium and thus indirectly show the presence of labelled DNA.

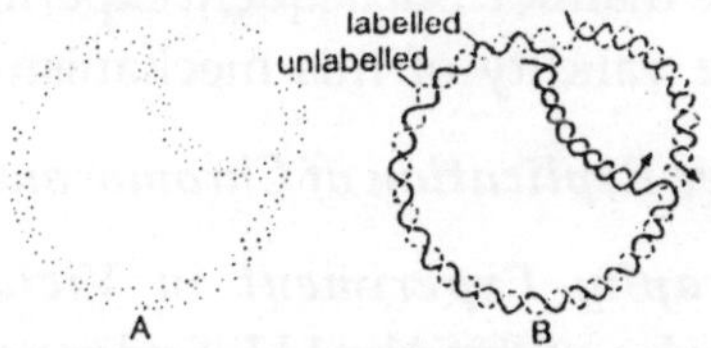

Cairns' autoradiography experiment, showing (A) an autoradiograph (drawing) and (B) its interpretation in terms of DNA.

Using the above technique, replication of DNA could be easily followed by J. Cairns and the results were reported in 1963. During incorporation of tritiated thymidine, autoradiographs could be prepared at regular known intervals. Autoradiographs from this replicating material prepared at regular known intervals demonstrated semi-conservative mode of replication. Lighter density of dots was considered to indicate that only one of the two strands was labelled, while a heavier density of dots indicated that both strands were labelled. Such a situation was actually observed.

The rate at which the replication proceeds could also be worked out by measuring the length of DNA undergoing replication in a known interval of time. The generation time was worked out by Cairns as 30 minutes in *E. coli.* The length of the chromosome was worked out to about 1 mm. The rate of replication thus would be approximately 30µ-40µ per minute (1 mm - 1000µ).

Figure indicates that after replication, one of the two strands in the daughter DNA molecules is derived from the parent molecule and the other is synthesized afresh. In θ shaped figure, which is obtained in the second cycle of replication in presence of label, two arcs in the split region would never be equally labelled. For instance, one arc would be twice as heavily labelled as the other arc. This is what was actually observed by Cairns. The observations thus clearly supported the semi-conservative nature of replication. In the above set of two experiments, it was demonstrated that DNA replicates in a semi-conservative manner. Subsequent experiments of different kinds proved the validity of this mechanism.

Semi-conservative Replication of Chromosomes in Eukaryotes

Autoradiography Experiment in Vicia Faba : Autoradiography was also utilized by J.H. Taylor and his co-workers for the study of duplicating chromosomes in the root tip cells of *Vicia faba*. Results of such experiments were first published in 1957. After incorporation of tritiated thymidine, when root tips were transferred to unlabelled medium, in the first generation of duplication both chromatids were labelled (interpreted as one DNA double helix in each chromatid, and only one of the two strands labelled). In the second cycle of duplication, in each chromosome, one of the two chromatids was found to be labelled. This was interpreted as showing semi-conservative mode of duplication.

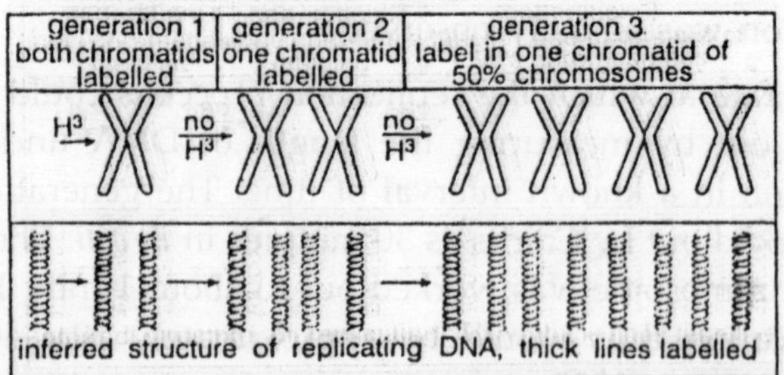

Tayor's experiment on Vicia faba root tips using autoradiography technique.

Taylor's experiment is important in so far as it demonstrated the semi-conservative mode of replication in chromosomes of a higher plant. But it should not necessarily mean semi-conservative mode of replication of DNA, because the organization of DNA in chromosomes of higher plants and animals is rather complex and not as simple as envisaged by Taylor, while interpreting his results.

Use of Harlequin Chromosomes : Semi-conservative replication of chromosomes can also be visualized through an examination of chromosomes that are allowed two rounds of replication in a medium containing bromodeoxy-uridine. These replicated chromosomes are then stained with a fluorescent dye and Giemsa stain to produce what are described as *harlequin chromosomes* (Fig.). The newly synthesized DNA stains differently than the parental DNA, so that the harlequin chromosomes have been found suitable for a study of (i) semi-conservative replication, (ii) crossing over and (iii) sister chromatid exchanges.

Semi-discontinuous DNA Replication : A. Kornberg initially characterized an enzyme, now called DNA polymerase I and

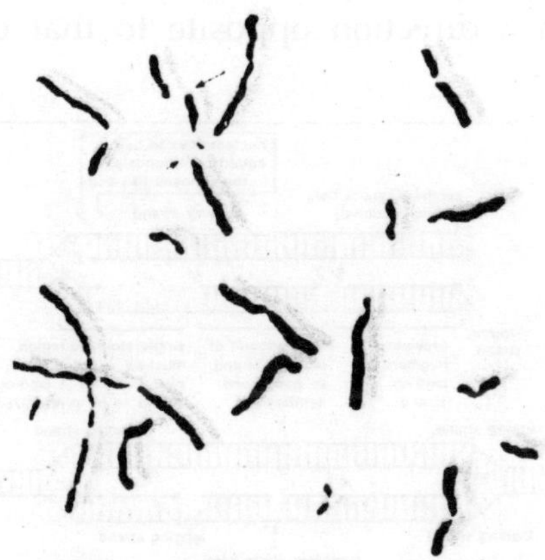

Harlequin chromosomes in a Chinese Hamster Ovary (CHO) cell. Arrow indicates sister chromatids exchange in a chromosome (see text for further details; from S. Wolff and J. Bodycote).

believed it to be responsible for DNA replication. However, later it was shown that this enzyme is mainly involved in DNA repair and not in DNA replication. Another enzyme DNA polymerase III is now known to be responsible for DNA replication and synthesizes DNA in 5' → 3' direction. Since the two strands of DNA have opposite polarities, DNA synthesis can not proceed on both strands, utilizing same enzyme, unless the synthesis proceeds in pieces. Such pieces called Okazaki pieces (after the name of discoverer) have actually been observed and it is established that DNA synthesis is discontinuous on one of the two strands and the segments later get fused with the help of ligase enzyme (Fig.).

At one time, it was felt that there was enough evidence to suggest that DNA synthesis is discontinuous on both the strands. However, now it is known that DNA synthesis is continuous on one strand and discontinuous on the other strand (Fig.). When the double helix of DNA unwinds, DNA replication on one of the two strands (3' to 5' strand) can easily proceed continuously in 5' to 3' direction. This is the leading strand. On the other strand (5'—3' strand), if the synthesis has to take place continuously in 5' to 3' direction, it has to be synthesized in a direction opposite to that on the leading strand.

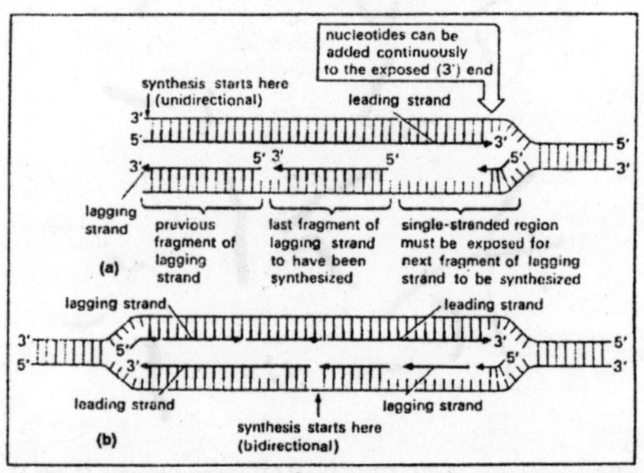

Semi-discontinuous DNA replication under conditions of (A) unidirectional replication, (B) bidirectional replication.

This strand is the lagging strand and on this strand, synthesis takes place in segments discontinuously and these segments are then fused to create an intact lagging strand. This behaviour where the leading strand is synthesized continuously and the lagging strand is synthesized discontinuously is called semi-discontinuous replication.

In certain viruses like adenovirus or (ϕ 29, linear DNA replicates from the two ends by strand displacement, so that both strands can be copied in 5' to 3' direction simultaneously without any need for discontinuous replication (Fig.). Therefore, it is obvious that semi-discontinuous replication is the result of a need to synthesize both strands simultaneously from the same origin.

Unidirectional and Bidirectional DNA Replication : John Cairns, from his autoradiography experiments which are described earlier in this chapter, had concluded that DNA synthesis starts at a fixed point on the chromosome and proceeds in one direction. It was subsequently realized that Cairns' results could be interpreted in terms of bidirectional replication also.

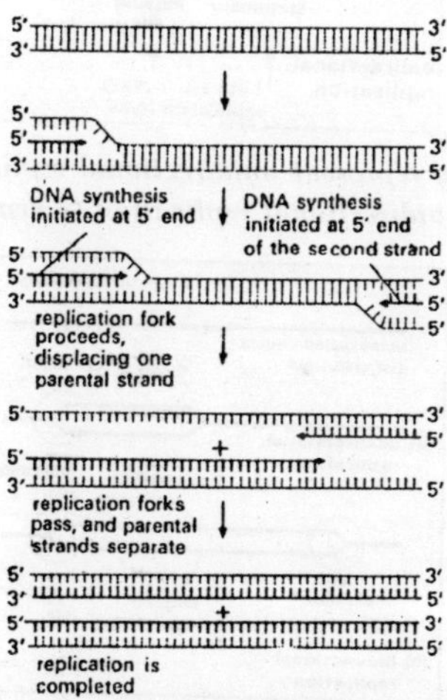

Continuous DNA replication on both strands initiated independently at the two ends (as observed in adenovirus).

Additional evidences supporting bidirectional replication were obtained from (i) autoradiography, (ii) electron microscopy and (iii) genetic studies. Evidence of bidirectional replication is also available in eukaryotes like yeast, fruitfly and mammals, where DNA synthesis starts at multiple sites, which start as loops and can be seen as expanding bubbles or eyes in electron micrographs (Fig.). Number of eyes within an initiation zone (5000-55000 bp) indicates the number of initiation sites for DNA replication (each about 100-500 bases).

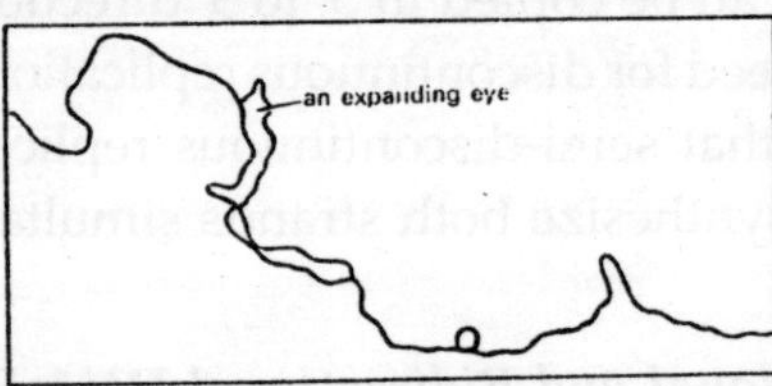

A diagrammatic representation of an electron micrograph of replicating eukaryotic DNA, showing formation of 'bubble' or 'eye'.

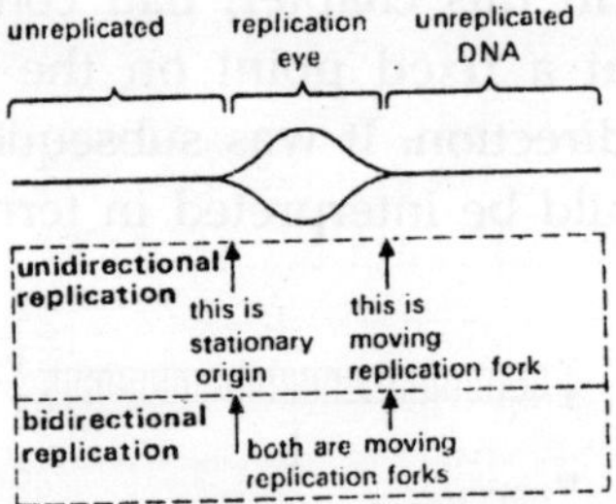

A replication eye may represent unidirectional replication (when only one end moves) or a bidirectional replication (when both ends move).

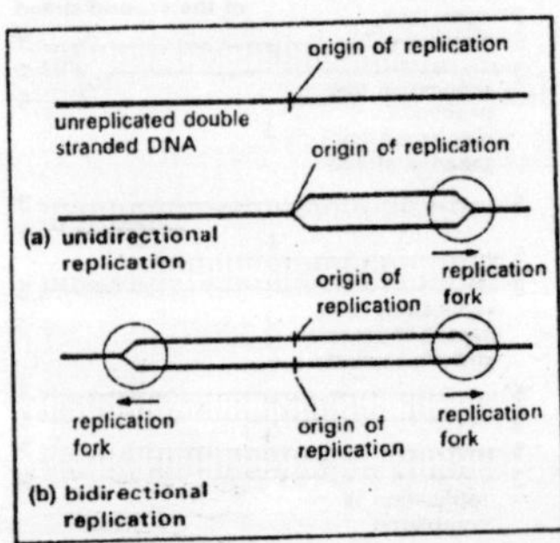

The difference between unidirectional and bidirectional replications in the movement of replication forks.

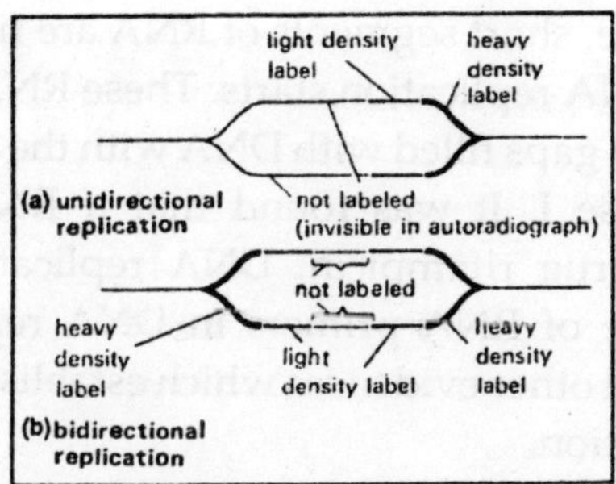

Distinction between unidirectional and bidirectional replications as revealed by radioactive labelling (note increasing densities of labelling on only one side or on both the sides of origin of replication).

In view of the above evidences, it is now known that DNA replication can be unidirectional or bidirectional, depending upon whether the replication from the point of origin proceeds only in one direction or proceeds in both the directions (Fig.). A replication eye may appear in both the situations, unless the replication starts from one of the two ends of a linear DNA molecule. However, in unidirectional replication, one of the two ends of the replication eye will be stationary and the other end will move with replication (Fig.). On the other hand, in bidirectional replication, none of the two ends will be stationary and both will be moving (Fig.). If radioactively labelled nucleotides are used during DNA synthesis, distinction between undirectional and bidirectional replication can be made. This can be done by the study of autoradiographs prepared after DNA synthesis has already proceeded for some time with the use of labelled nucleotides. Labelling on both the replication forks will indicate bidirectional replication and labelling only at one fork will suggest unidirectional replication (Fig.). An example of unidirectional replication is the replication of mitochondria! DNA (mtDNA) by D-loops in vertebrates.

RNA Primers in DNA Replication

No DNA polymerase enzyme is known, which can initiate synthesis of DNA chain without the availability of a primer strand. Therefore, short segments of RNA are first synthesized before actual DNA replication starts. These RNA segments are later excised and gaps filled with DNA with the help of enzyme DNA polymerase I. It was found that if RNA synthesis is stopped by a drug rifampicin, DNA replication stops thus proving the role of RNA primers in DNA replication (Fig.). There are several other evidences which established this feature of DNA replication.

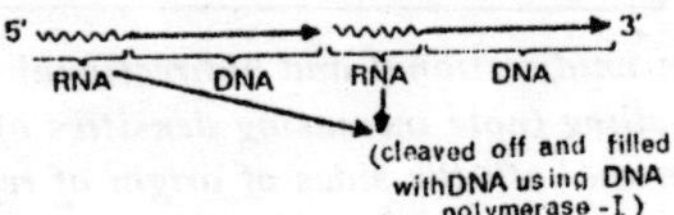

Discontinuous DNA replication using RNA primers at 5′ ends of segments.

Regulation of DNA Replication by Anti-sense RNA Primers: While studying DNA replication in an *E. coli* plasmid (ColEl), Jun-ichi Tomizawa in USA demonstrated that certain RNA molecules, synthesized on DNA strand complementary to the one that produces RNA primers, are actually synthesized. These were described as antisense inhibitor molecules and the rate of multiplication of the plasmid depended on the ratio of RNA primers to these antisense molecules. Obviously antisense RNA molecules hybridize with RNA primers and render them incapable of initiating DNA replication.

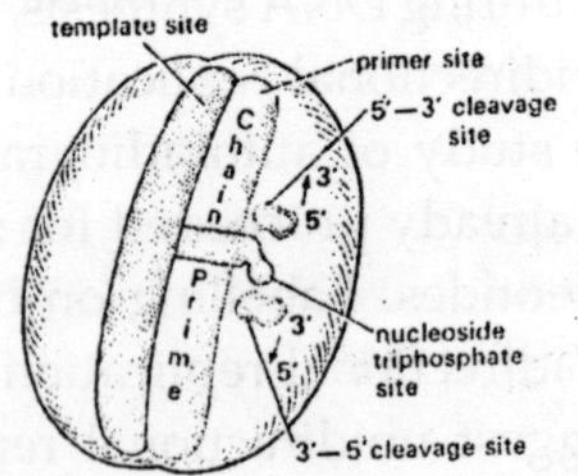

A model of DNA polymerase-I enzyme, showing five different sites, (redrawn from Kornberg, DNA synthesis—1974).

Antisense RNA is also used for regulation of protein synthesis (gene expression) and has been shown to offer useful technology for therapeutic uses against certain diseases caused by viruses, etc.

Prokaryotic DNA Polymerases

Three different prokaryotic DNA polymerases are known, of which DNA polymerases I and II are' meant for DNA repair and DNA polymerase III is meant for actual DNA replication, (i) DNA polymerase I (isolated around 1960 by Arthur Kornberg) was the first enzyme suggested to be involved in DNA replication. This enzyme is now considered to be a DNA repair enzyme rather than a replication enzyme. It has five active sites shown in Figure and its properties are shown in Table. This enzyme is mainly involved in removing RNA primers from Okazaki fragments and fills up the gaps due to its 5'→3' polymerising activity. However, a ligase enzyme is needed for covalent bonding after the gap is filled by DNA polymerase I. It can also remove thymine dimers produced due to UV irradiation and fill the gap due to excision. Polymerising activity of DNA polymerase I is shown in Figure and an exonuclease activity removing mispaired nucleotide is shown in Figure.

A Comparison of Properties of DNA Polymerases of E. coli.

Property	*DNA polymerases*		
	I	*II*	*III*
1. Polymerisation 5' → 3'	yes	yes	yes
2. Exonuclease 3' → 5'	yes	yes	yes
3. Exonuclease 5' → 3'	yes	no	no
4. Use of primed single strands	yes	no	no
5. Use of nicked duplex	yes	no	no
6. Molecular weight	109,000 (single chain)	120,000 (single chain)	>250,000 (heleromultimetric chain)
7. Molecules per cell	400	not known	10-12
8. Gene	*polA*	*polB*	*polC (dnaE), dnaN, dnaX, dnaQ, dnaT*
9. Nucleotides polymerised at 37°, molecules/minute	Up to 1,000	Up to 50	Up to 15,000
10. Affinity for nucleoside triphosphate (NTPs) precursors	Low	Low	High

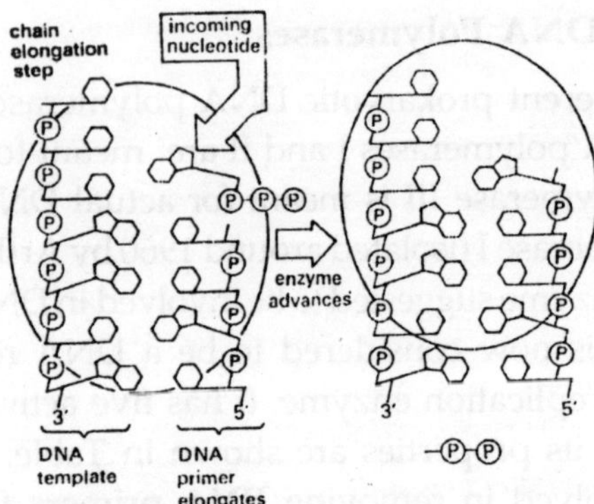

A model showing polymerization of nucleotides, on the primer site of DNA polyinerase-I (redrawn from Koinberg, DNA synthesis—1974).

This is described as proofreading function of this enzyme. DNA polymerase I consists of two fragments: a larger fragment, called Klenow fragment (38kD), which contains 3'—5' exonuclease activity with 5'—3' polymerising activity and a smaller fragment (35kD), which contains 5'—3' exonuclease activity. The activity of this enzyme (due to Klenow fragment) in using nicked DNA *in vitro* is unique and is utilized to label DNA molecules with radioactive nucleotides. This is called nick translation, and is shown in Figure. DNA polymerase I enzyme is synthesized under the control of gene *polA* located on *E. coli* map at a position of 75 minutes, (ii) DNA polymerase II (isolated in 1970-71) resembles DNA polymerase I in its activity to bring about the growth in 5' $\rightarrow$ 3' direction, using free 3'—OH groups, but mainly uses duplexes with short gaps only. It can not use nicked duplexes (unlike DNA polymerase I).

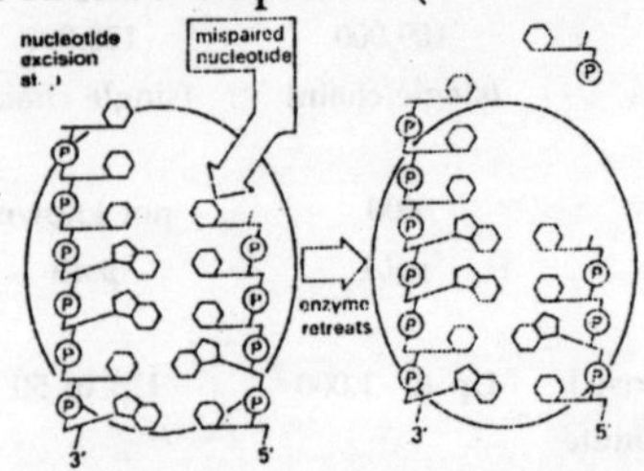

Removal of a mispaired nucleotide by exonuclease activity of DNA polymerase I.

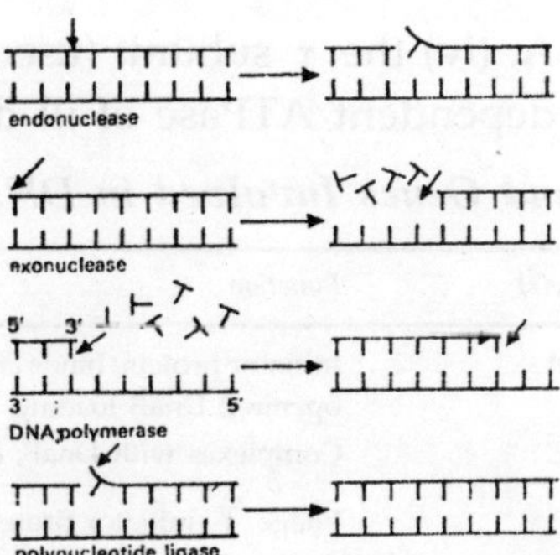

Effect of four classes of enzymes on DNA duplex, showing cleavage in case of nucleases (endonuclease and exonuclease) and repair in the other two cases (redrawn from Watson,* Molecular Biology of the Gene*—1987).

Although it has 3' → 5' exonuclease activity, it lacks 5' → 3' exonuclease activity (exonuclease activity means cleavage of nucleotides only at the end, while endonuclease breaks DNA strand at an internal position). Since *polB*$^-$ mutants lacking DNA polymerase II appear normal in growth and conduct DNA replication normally, this enzyme can not be a replication enzyme. DNA polymerase II enzyme is also involved in DNA repair, (iii) DNA polymerase III holoenzyme (isolated in 1972) plays an essential role in DNA replication. There is intracellular scarcity of this enzyme (only 10-20 molecules per cell), thus requiring rapid recycling.

It is a heteromultimeric enzyme with ten units. All the ten subunits, listed in Table, are needed for DNA replication *in vitro*. The ten subunits of holoenzyme are constituted into the following four components: (i) a polymerase 'core' enzyme consisting of one molecule each of subunits α , ε and θ, and having the ability to synthesize DNA; (ii) a sliding clamp, consisting of a dimer (β_2), which, binds to DNA in an ATP dependent reaction catalyzed by 'γ complex' to form the pre-initiation complex (PIC) ; (iii) 'γ complex' consisting of six polypeptides with five subunits $(\gamma_2\delta_1\delta_1\chi_1\Psi_1)$; it is also called matchmaker, because it helps in the loading of sliding clamp

(β_2) onto the DNA. (iv) the τ subunit (used as a dimer, τ_2), which is a DNA dependent ATPase of ill defined function.

Proteins and Genes Involved in DNA Replication

Protein	*Gene(s)*	*Function*
DnaA	*dnaA*	Initiator protein; binds *oriC;* promotes double helix opening; DnaB loading
DnaC	*dnaC*	Complexes with DnaB; delivers DnaB to DNA
λ O	O	Phage γ initiator protein; binds ori γ ; promotes double helix opening; complexes with λ P and DnaK
λ P	*P*	Complexes with DnaB; delivers DnaB to oriX complex
PI RepA	*repA*	Plasmid PI initiator protein; binds *oriPl;* complexes with DnaJ
IHF	*himA, hip*	Sequence-specific double-strand DNA binding protein; 'binds to *oriC*
FIS	*fis*	Sequence-specific double-strand DNA binding protein; binds to *oriC*
HU	*hupA, hupB*	Non-specific double-strand DNA binding protein
RNA polymerase	*rpoA, rpoB, rpoC, rpo*	Transcriptional activation of *oriC* and *ori* λ , leading strand Dpriming of ColEl
DnaK	*dnaK*	Heat shock protein; disassembly of ori λ complex; monomerization of P1 RepA
DnaJ	*dnaJ*	Heat shock protein; tagging of proteins for action by DnaK
GrpE	*tfrpE*	Heat shock protein; complexes with DnaK
DnaB	*dnaB*	5' to 3' helicase and activator of primase
Primase	*dnaG*	Primer synthesis
SSB	*ssb*	Binds single-stranded DNA
PriA	*priA*	Assembly of the ϕ X-type primosome at *pas;* 3' to 5' helicase; (*n'* or factor Y)
PriC	*priC*	Assembly of ϕ X-type primosome (*n"* protein)
DnaT	*dnaT*	Assembly of ϕ X-type primosome (protein i or factor X)
DNA polymerase III holoenzyme		
α	*dnaE*	Polymerase — Core
ε	*dnaQ*	3' to 5' exonuclease

Contd...

Protein	*Gene(s)*	*Function*
θ	*holE*	Stimulates ε PollII
τ	*dnaX*	Dimerization of core, ATPase
γ	*dnaX*	Binds ATP Pol III Poll III
δ	holA	Binds t P
δ'	*hdlB*	Cofactor for γ ATPase and stimulates clamp loading γ complex'
χ	*holC*	Binds SSB
ψ	*hnlD*	Bridge between x and 7
β	*dnaN*	Processivity factor; 'sliding clamp'
Poll	*polA*	Primer removal and gap filling; initial leading strand synthesis of ColEl
RNase H	*rnh*	Primer removal, primer processing for leading strand synthesis of ColEl
Ligase	*tig*	Joins nascent DNA fragments
Gyrase	*gyA, gyrB*	Type II topoisomerase; replication swivel, DNA supercoiling, decatenation
Topo IV	*purC, parE*	Decatenation; presumptive type II topoisomerase
TBP	*tus*	Ter binding protein; stalling of replication forks

During initiation of DNA synthesis, sliding clamp is first loaded onto DNA in an ATP dependent reaction catalyzed by matchmaker 'γ complex' to form the pre-initiation complex. This is followed by the assembly of the core enzyme with the pre-initiation complex to form the initiation complex in an ATP independent reaction, during which, the matchmaker (γ complex) dissociates. The sliding clamp confers processivity (tendency to remain on a single template rather than to dissociate and reassociate again) to core enzyme and is passively pulled along with core during polymerization. The slilding clamp (β_2) is a ring shaped, head to tail, dimer with a central cavity of sufficient diameter to accommodate duplex DNA. It is a homologue of eukaryotic PCNA (see later).

DNA pol holoenzyme (HE) has an asymmetric structure, consisting of 16 polypeptides involving 10 subunits as follows : β_2 $(\alpha\ \varepsilon\ \theta)_2 \tau_2 (\gamma_2 \delta_1 \delta'_1\ \chi_1 \Psi_1)$.

The subunits of holoenzyme assemble in an orderly fashion as shown in Figure. Following sub-assemblies of the holoenzyme have also been recognized :

(i) Pol III' consists of two cores and a dimer of $\tau\,(\tau_2)$, so that the polymerase acts in pair for coordinated replication of both strands of a duplex.

(ii) Pol III* assembly contains 14 polypeptides $\{(\alpha\ \varepsilon\ \theta)_2 \tau_2\ (\gamma_2 \delta_1 \delta'_1\ \chi_1 \gamma_1)\}$ involving 9 subunits and lacks the sliding clamp or β_2, which confers high speed and processivity to polymerases.

Table shows a comparison of DNA polymerase III with DNA polymerase I and DNA polymerase II of *E. coli.* A comparison of synthetic activity of DNA polymerases with other related enzymatic activities (endonuclease, exonuclease, ligase) is presented in Figure.

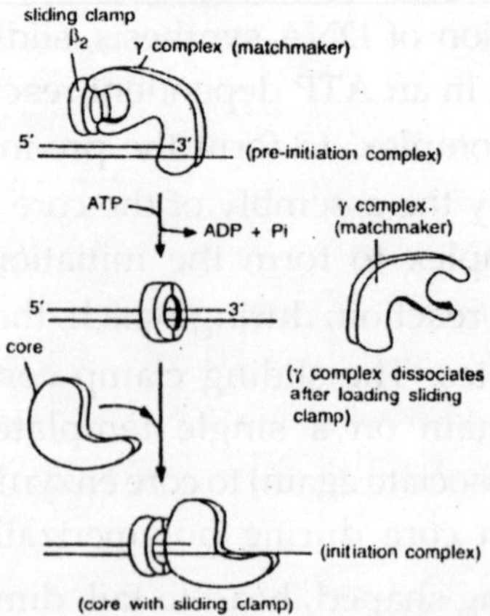

Formation of initiation complex' for DNA replication by the assembly of core enzyme (not the role of 'sliding clamp' and 'γ complex)

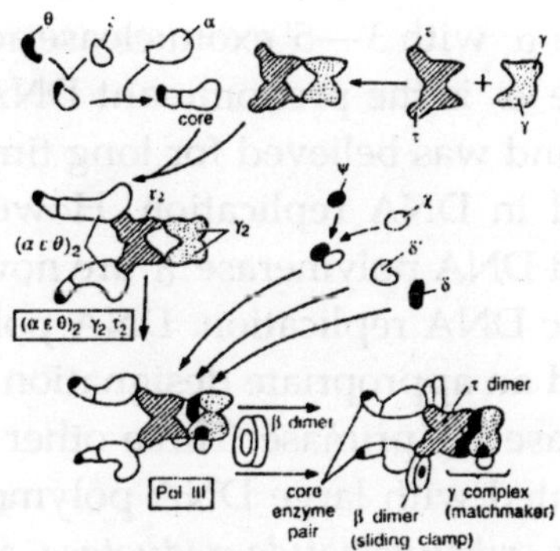

Assembly of DNA polymerase holoenzyme (HE) from its 16 subunits.

Eukaryotic DNA Polymerases

Five different eukaryotic DNA polymerases have been identified, from materials like yeast cells, rat liver, human cell cultures and tumour cells. These include the following : (i) DNA polymerase α, a relatively high molecular weight enzyme, also called *cytoplasmic polymerase* (also appropriately called *large polymerase*, since it is also found in the nucleus and since another small nuclear polymerase is also available in cytoplasm); (ii) DNA polymerase β or *nuclear polymerase* (also called *small polymerase)*, found only in vertebrates, and not in budding yeast, nor in other lower eukaryotes; (iii) DNA polymerase γ or *mitochondrial polymerase*, encoded in the nucleus; (iv) DNA polymerase δ, another novel enzyme reported from mammalian cells, is PCNA dependent for DNA-synthetic processivity (PCNA = proliferating cell nuclear antigen; processivity = property of continuous synthesis); (v) DNA polymerase ε also purified and reported from mammalian HeLa cells and budding yeast, is PCNA independent.

In some other cases, a distinction between DNA polymerase δ and DNA polymerase e has not been possible. Another calf thymus enzyme designated as DNA polymerase δ I represented

DNA polymerase α with 3'—5' exonuclease activity. The large DNA polymerase α is the predominant DNA polymerase in eukaryotic cells and was believed for long time to be the only enzyme involved in DNA replication. However, both DNA polymerase a and DNA polymerase δ are now known to take part in eukaryotic DNA replication. DNA polymerase α has four subunits and an appropriate designation for this enzyme is DNA polymerase α /primase. Three other enzymes which are found associated with large DNA polymerase α include *thymidine kinase, ribonucleotide reductase* and *thymidylate synthetase.* They are also believed to be involved in DNA replication.

Nuclear small DNA polymerase β resembles large DNA polymerase α in its replicative activity, but differs in some other respects. Its quantity remains unchanged, while that of large polymerase changes with need. It is speculated that small enzyme may be involved in repair and recombination. The differences between these polymerases are shown in Table.

Eukaryotic DNA Polymerases : Protein Structure, Cellular Location, and Properties (Wang, 1991)

	DNA polymerase				
	α (*Pol I**)	β	γ	δ (*Pol II**)	ε (*Pol III**)
Protein Structure					
catalytic polypeptide (kDa)	165[a]	40	140	125	255
associated subunits (kDa)	70 58 48	None	Unknown	48	Unknown
Yeast gene	*POLI* (*CDCI7*)	None	*MIP1*	*POL3* (*CDC2*)	*POL2*
Cellular location	Nuclear	Nuclear	Mito	Nuclear	Nuclear
Associated Enzyme Activities					
5'-3' exonuclease	None	None	None	None	None
3'-5' exonuclease	None[b]	None	Yes	Yes	Yes
primase	Yes	None	None	None	None

Contd...

	DNA polymerase				
	α (Pol I*)	β	γ	δ (Pol II*)	ε (Pol III*)
Properties					
processivity					
inherent	Med	Low	High	Low	High
with PCNA	—	—	—	High	—
fidelity	High	Low	High	High	High
Inhibitors					
aphidicolin	Yes	No	No	Yes	Yes
N-ethylinaleimide	Yes	No	Yes	Yes	Unknown
butylphenyl-dGTP	Yes	No	No	No	No
dideoxynucleoside 5'-triphosphate (ddNTP)	No	No	Med	No	No
carbonyl-diphosphonate	No	Un-known	Un-known	Yes	Med

* *earlier designation*

[a]molecular mass deduced from primary protein sequence.

[b]cryptic exonuclease in the isolated catalytic polypeptide was reported in Drosophila melanogaster embryo.

Replicons for DNA Replication

DNA replication in prokaryotes as well as in eukaryotes is achieved in discrete units called replicons, which may vary in a genome from one in bacteria like *E. coli,* and 500 in yeast to several thousands in animals and plants (Table). In *E. coli* there is a single replicon with the origin, identified as a genetic locus *ori*C (245 bp). The sequence of this locus has been isolated and cloned in plasmids and allows autonomous replication of plasm id DNA just like bacterial chromosome. Isolated and cloned telomeres and centromeres from yeast are also similarly used for preparing yeast artificial chromosomes (YACs) The origins of replication have been identified in bacteria, yeast, chloroplasts and mitochondria. These are A:T rich, a feature which is related to unwinding of DNA to initiate replication. In *E. coli* there are also termination sites *(ter A-F)*, each consisting of ~ 23 bp. Termination of DNA replication requires product of *tus* gene *(Tus* protein or ter binding protein = TBP), which recognizes *ter* sites.

In a eukaryotic genome, there are many replicons, an individual replicon being small and the rate of replication being much slower. Not all replicons start replication at the same time, but the initiation of replication at different replicons takes place in an ordered manner. There is no evidence of termination sites in eukaryotes (unlike *E. coli)*, but termination is brought about when a termination fork meets another fork proceeding towards it from origin of an adjoining replicon.

Prokaryotic and Eukaryotic Replicons

Organism	*No. of replicons*	*Average length (kb)*	*Fork movement (bp/min)*
Bacteria *(E. coli)*	1	4200	50,000
Yeast (*S. cerevisiae)*	500	40	3,600
Fruitfly *(D. melanogaster)*	3500	40	2,600
Toad (X. *laevis)*	15000	200	500
Mouse *(M. musculus)*	25000	150	2,200
Bean *(Vicia faba)*	35000	300	not known

DNA Replication in Prokaryotes

Among prokaryotes, DNA replication *in vitro* has been studied in great detail in *E. coli* and its phages and plasmids. The techniques of genetics and biochemistry have been used to identify genes and their products (proteins) involved in DNA replication (Table). The most extensively studied examples include DNA replication from (i) *E. coli* chromosomal origin *oriC;* (ii) phage replication origin, *oriX;* (iii) phage PI lysogenic origin *oriPl;* and (iv) ColEl and pBR322 plasmid origin. Among these four systems, more variation is found in the initiation than in the elongation and termination of DNA replication.

Experimental Approaches for the Study of DNA replication : Three different approaches have been used for the study of DNA replication in prokaryotes : (i) Isolation of mutants. Replication defective conditional mutants have been selected, which are, temperature sensitive dna_{ts} (non-conditional mutants, inhibiting DNA synthesis are usually lethal and are therefore not useful for a study of DNA replication).

DNA synthesis in these mutants stops, when shifted to non-permissive temperature (*e.g.* 45°C). These can be either the replication initiation/termination mutants (also called 'slow stop mutants') which do not stop DNA synthesis immediately after shifting to non-permissive temperature (*e.g. dnaA, dnaB* and *dnaC),* or may be replication extension/elongation mutants (also called 'quick stop mutants'), which stop DNA synthesis immediately after shifting (*e.g. dnaB, dnaC, dnaE, dnaG, dnaN* and *dnaX).*

Thus the proteins synthesized by the former class must be involved in initiation/termination and those synthesized by the latter must be involved in elongation: In former class, DNA synthesis does not stop immediately, because DNA synthesis continues for one complete round on replication units where DNA synthesis was already initiated. Some of the viable mutants, which are not temperature sensitive, but are defective in DNA replication, have also been found useful, (ii) Use of inhibitors. Sensitivity of the initiation of DNA synthesis to rifampicin suggested that RNA synthesis is involved in the initiation of DNA replication. 'Novobiocin' and 'nalidixic acid' inhibited elongation by inhibiting 'DNA gyrase', which functions as a replication swivel, (iii) Isolation of proteins. Several proteins, which have a role in DNA synthesis, have been identified, by either their ability to complement the defective mutants or by their requirement for *in vitro* replication of plasmid or phage DNA. Genes for several of these proteins have now been isolated and cloned.

Initiation of DNA Replication

Initiation of Replication in Double Stranded DNA : A comparison of the initiation pathways used at *oriC, ori* λ , *oriP1* and *ColE1* (studied *in vitro)* suggests three steps :

(i) recognition of the origin;

(ii) opening of duplex to generate a region of single stranded DNA and

(iii) capture of DnaB protein (a helicase).

The events for initiation are sometimes also classified into 'preprinting' (which occurs only at the origin), 'priming' (which recurs with the initiation of each Okazaki fragment during elongation phase) and the formation of pre-initiation and initiation complexes. There are variations in different cases, in the mechanism of initiation, and also in the proteins involved in initiation. Only *E. coli ori*C will be described here, and differences will be indicated, wherever necessary and possible.

Preprinting and Priming : For initiation from *oriC,* following steps are involved (as deduced from *in vitro* studies) : (i) *DnaA-ATP complex* (10-20 monomers of DnaA) binds at 9bp inverted repeat regions *(R_1, R_2, R_3, R_4)* of *oriC* and promotes opening of the DNA in a region of three direct repeats of a 13-bp sequence (called *13-mers;* see Fig.).

The opening occurs from right *13-mer* leftwards, and requires negatively supercoiled DNA and Hu or IHE proteins, besides DnaA-ATP complex, (ii) DnaB is transferred to exposed single stranded DNA, with the help of DnaC, in presence of ATP. DnaC, perhaps, recognizes DnaA and thus helps in the transfer of DnaB, but DnaC itself is released from DnaB during the transfer reaction, (iii) 'DnaB helicase' causes unwinding of the DNA, on addition of ATP, SSB (single strand binding protein) and DNA gyrase. This leads to the formation of a preprinting complex.

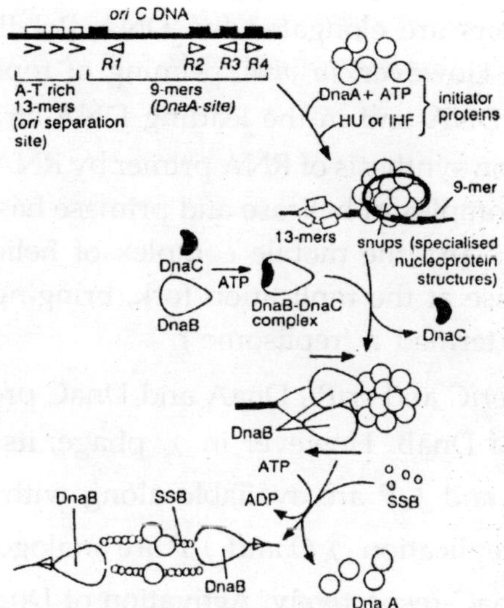

Model of the initiation of DNA replication at oriC of E. coli.

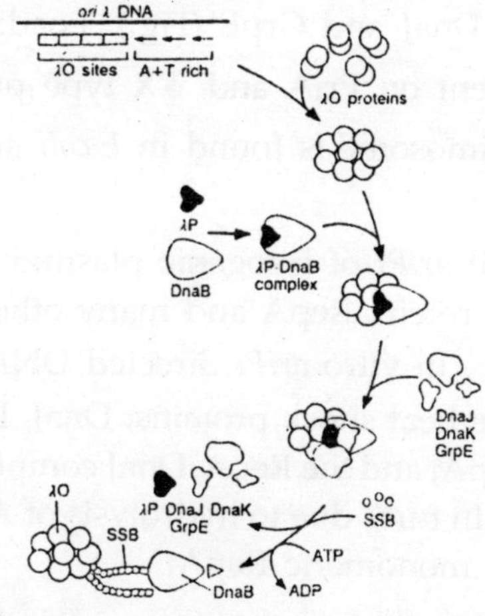

Model of the initiation of DNA replication at ori λ of phage λ.

Unwinding and replication from *oriC* proceeds in both directions (bidirectional), SSB binding occurs on single stranded regions, and two DnaB complexes, are loaded one on each strand, (iv) Unwinding is followed by the synthesis of RNA primers by DnaG primase (through its interaction with DnaB) and the primers are elongated by 'DNA Pol III HE' (HE = holoenzyme). However, *in vivo,* priming of replication, in *E. coli* DNA, λ DNA and in the leading DNA strand of ColEl, is dependent on synthesis of RNA primer by RNA polymerase. (The mobile complex of helicase and primase has been termed a 'primosome' and the mobile complex of helicase, primase and polymerase at the replication fork, bringing about DNA replication is termed a 'replisome').

On both, *oriC* and *oriPI,* DnaA and DnaC proteins help in the transfer of DnaB. However in λ phage, its own specific proteins λ *O* and λ *P* are available along with many *E. coli* proteins, for replication. λ *O* and λ *P* are analogous to *E. coli's,* DnaA and DnaC respectively. Activation of DnaB in λ phage requires release of λ P, which is assisted by three heat-shock proteins, DnaK, DnaJ and GrpE (Fig.). Loading of DnaB on ColEl is dependent on PriA and ϕX type primosome. (The other type of primosome is found in *E.coli* and is known as *OriC* type).

Replication at *oriPI* of lysogenic plasmid P1 requires the phage-encoded protein, RepA and many other proteins used by *oriC* and *ori*λ. In vitro *oriPl* directed DNA synthesis also depends on three heat shock proteins, DnaJ, Dnak and GrpE. DnaJ binds to RepA, and the RepA-DnaJ complex is recognized by DnaK, which in turn, due to hydrolysis of ATP, triggers the release of active monomeric RepA.

The role of GrpE in activation of RepA is not clearly understood. Once active in monomeric form, RepA binds to *oriPI,* so that *RepA-oriPl* DNA complex is a substrate for DNA replication. DnaA binds to its sites and leads to unwinding of DNA. DnaB helicase is now transferred to the origin, (perhaps through DnaC-DnaA interaction), and activation of DnaB helicase (without the involvement of DnaJ, Dnak and GrpE) leads to unwinding of the DNA. Now primase synthesizes primers, and DNA pol III HE catalyzes DNA synthesis (Fig.). *In vitro,* the replication at *oriPI* is unidirectional.

Initiation of DNA replication from ColEl origin is catalyzed by *E. coli* proteins, but differs in several essential features. On the leading strand RNA polymerase synthesizes the transcript forming RNA-DNA hybrid, that is processed by RNase H generating 3' OH group, which serves as a primer for elongation by DNA polymerase I forming DNA-DNA hybrid and displacing the other strand of the plasmid DNA. The displaced strand exposes a site known as 'primosome assembly site' or *pas.* A protein-DNA complex, called 'ϕX-type primosome' (including DnaB) is assembled at this site and is recognized by DnaG primase for initiation of primer syrthesis. Thus the primer synthesis in ColEl differs on two strands and the replication is unidirectional. Eventually DNA pol III HE replaces DNA pol I on the leading strand and DnaG primase on the lagging strand for DNA elongation (Fig.).

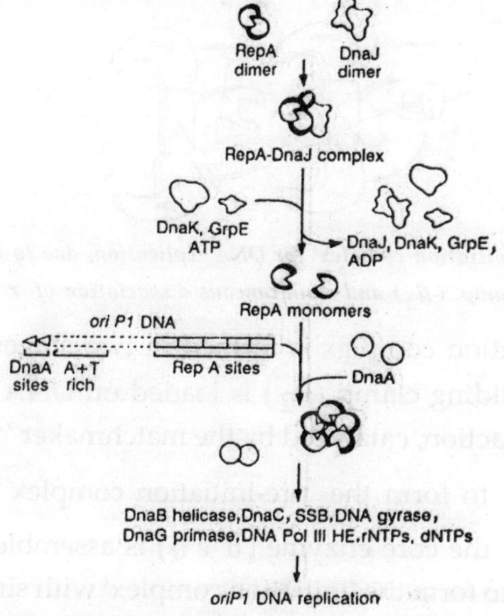

Model of the initiation of DNA replication or oriP1 of phage P1

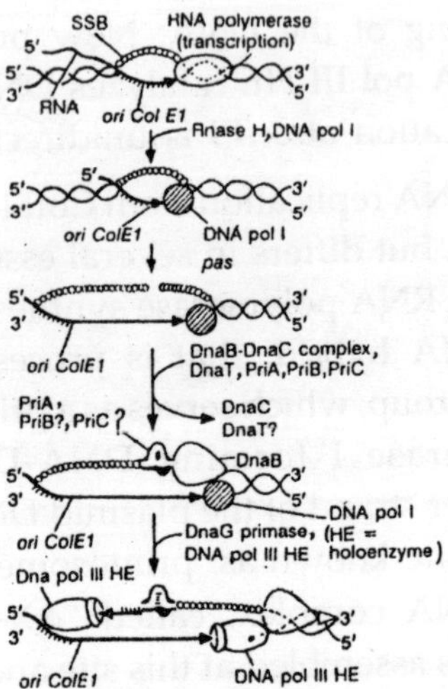

Model of the initiation of DNA replication at ColE1/BR322 origin.

Formation of Pre-initiation and Initiation Complexes : As mentioned earlier, after prepriming and priming, an initiation complex for DNA-synthesis is formed utilizing the components of DNA polIII HE.

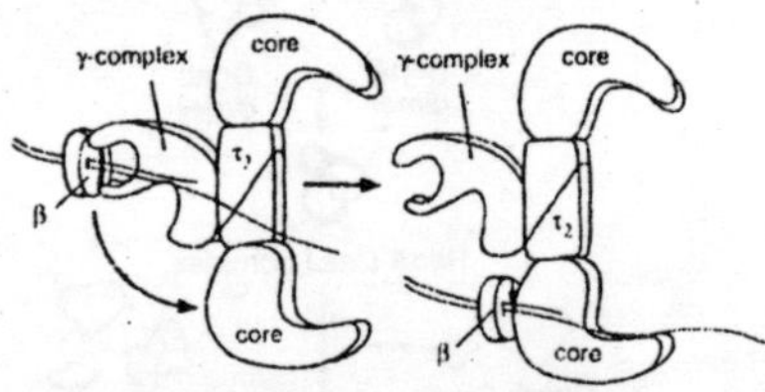

Formation of 'initiation complex' for DNA replication, due to repositioning of 'sliding clamp' (β_2) and simultaneous dissociation of γ complex'.

The initiation complex is formed in two stages: (i) In the first stage, slilding clamp (β_2) is loaded on DNA in an ATP dependent reaction, catalyzed by the matchmaker 'γ complex' $\left(\gamma_2\delta_1\delta,.\chi_1\psi_1\right)$ to form the 'pre-initiation complex.' (ii) In the second stage, the core enzyme ($\alpha\ \varepsilon\ \theta$) is assembled with β_2 bound DNA to form the 'initiation complex' with simultaneous dissociation of 'γ complex' from the sliding clamp (β_2) (Fig.).

The 'γ complex' though released from the sliding clamp (β_2), still remains associated with core ($\alpha\ \varepsilon\ \theta$) and β_2 on DNA through τ_2. In the holoenzyme, τ_2 acts as a bridge between 'core' and 'γ complex' to hold them together. The 'sliding clamp' (β_2) with double stranded DNA is now repositioned in contact with core, but away from 'γ corriplex' (Fig.). The retention of 'γ complex' in association with the rest of the enzyme, DNA pol III HE (at the replication fork) helps in multiple initiation events on the lagging strand. The protein and enzyme complex associated with the initiation complex formed at the replication fork, is also described as 'replisome'.

Initiation of Replication in Single Stranded DNA Phages : In many cases (including phage M13, which is single stranded and mtDNA) it is the RNA polymerase, which is involved in priming the DNA syntheses through the synthesis of small RNA segment, which precedes DNA synthesis (as earlier discussed). However, in phage G4 and in phage $\phi \times 174$ (both single stranded DNA), in place of RNA polymerase, a much smaller protein called primase (coded by *dnaG* gene) is utilized. This primase enzyme synthesizes RNA from one unique specific site each in phages G4 and Ml3, but at multiple sites in $\phi\ \times 174$.

In $\phi \times$l74 (as also in G4 and M13), single stranded DNA is first coated with SSB protein, which is converted to a prepriming intermediate by six proteins named PriA, PriB, PriC, DnaT, DnaB and DnaC. The prepriming complex of six proteins assembles at a point described as 'primosomc assembly site' or "pas' (hairpin formed due to intrastrand pairing), and dnaG primase can now synthesize short RNA molecules at several sites. The complex of proteins including primase at the priming site is called primosome. Among the prepriming proteins PriA seems to be responsible for selecting the site (pas) for the prepriming complex and for displacing SSB. The other proteins (PriB, PriC) have an unknown prepriming function. It has, however, been shown that in the absence of SSB coating, primase in association with DnaB protein can synthesize RNA primer without the need of other five, proteins (PriA, PriB, PriC, DnaT and DnaC), but a prior coating of SSB would need these five proteins for the synthesis of RNA primer.

Elongation of DNA Chain: Replication Elongation Proteins : As discussed above, initiation of DNA replication involves loading and activation of DnaB helicase. DnaB directs other necessary enzymes (particularly DnaG primase and DNA pol III HE) to the template and help in DNA elongation. These elongation proteins and the mechanism involved in elongation will be briefly described in this section.

DnaB Protein : DnaB is the major helicase, although other helicases have been discovered. It occurs as hexamer and unwinds the template in 5'-3' direction during *oriC* replication *in vitro,* at a rate of 500-1000bp/sec. The direction of unwinding requires DnaB to be placed on lagging strand; ahead of the polymerase on the leading strand. It also activates primer synthesis by DnaG primase. Since a DNA helicase not only unwinds DNA, but also translocates along the DNA during replication, it is believed to be a 'motor protein', with which it has many similarities.

DnaG Primase : The primase (DnaG) synthesizes a unique oligonucleotide primer on SSB coated single stranded phage G4 DNA. The sequence recognized by primase is over l00nt (nt = nucleotides) long containing a secondary structure. Since primer synthesis requires activation by DnaB, the latter is often described as a 'mobile promoter' of primer synthesis.

DNA Polymerase III Holoenzyme : DNA pol III HE is the large multiprotein complex responsible for elongation of DNA replication in *E. coli.* It has at least 10 subunits, each with specific function (Table). The holoenzyme consists of a 'core polymerase' and' its 'accessory factors', which are analogous to the components of phage T4 and eukaryotic enzymes. In the holoenzyme, β subunit is twice as abundant as the other subunits, and in the cell its abundance is 300 times that of the polymerase assembly. It has a 'sliding clamp' mode of DNA binding. It has similarity with T4 *gene 45* protein and human PCNA.

Comparison of Pol III Holoenzyme with other Replicases: DNA polymerase III holoenzyme of *E. coli,* as shown above, consists of the following three components (with 10 subunits): (i) a polymerase (core), (ii) a sliding clamp (β_2) and (iii) a clamp loader (γ complex). These three components can be compared with corresponding components in replicases of eukaryotes (Table).

A Comparison of the Components of Replicases in E. coli, Eukaryotes and T4 Phage

Components of replicases	*E. coli*	*Eukaryotes*	*T4 phage*
Polymerase/ exonuclease	Core (3 subunits)	Pol δ (2 subunits)	g43p (1 subunit)
Clamp loader (matchmaker) subunits)	γ complex (5 subunits)	RF-C ,γ complex (5	g44/62p complex (2 subunits)
Sliding clamp	β_2	PCNA	g45p
Proteins interacting with clamp	Pol III, Pol II	Pol δ, & Pol,ε, others,	g43p (pol), RNA polymerase

Other Proteins for DNA Elongation : Other proteins involved in DNA elongation are (i) DNA gyrase (a topoisomerase), which removes the topological strain associated with DNA unwinding; (ii) SSB (single stranded DNA binding) protein, that binds tightly to single stranded DNA thus stabilizing the unwound DNA; (iii) RNase H, which removes RNA primers; (iv) DNA pol I, which is used for filling the gap created due to removal of RNA primers and (v) DNA ligase, which converts Okazaki fragments into a continuous strand.

Replication Fork Movement

During initiation to elongation transition, DnaB (while travelling in 5'-3' direction) generates a replication fork by opening the DNA duplex (Fig.). The strand having DnaB will thus be the lagging strand template. DNA primase associates with DnaB and synthesizes multiple primers for lagging strand and single primer for leading strand. The dimeric DNA pol III HE interacts simultaneously with both template strands. The leading strand polymerase moves in the same direction as DnaB, while copying the opposite strand. For the synthesis of lagging strand, the DNA polymerase will have to work on the same strand to which DnaB is bound, but it travels in opposite direction. The Tieltcase and primase activities may be provided by DnaB and primase or alternatively by ϕ X-type primosome.

DnaB, primase and 'DNA pol III HE' function together in elongation, although DnaB and DNA pol III HE seem to be tightly bound to the fork and stay bound throughout the reaction, an attribute described by saying that DnaB and DNA polymerase assembly is processive. In contrast, primase, 'β_2' and 'γ complex' act distributively (they dissociate and reassociate *i.e.* do not stay bound throughout the reaction).

Continuous Synthesis on Leading Strand : The leading strand is primed only once on each of the two parental strands in bidirectional replication. In bidirectional replication, the first primer synthesized can serve as the 'leading strand primer' and subsequent primers serve as 'lagging strand primers'. No special mechanism is, therefore, needed for priming of the leading strand in bidirectional replication, and the DNA polymerase complex on leading strand is extremely processive.

However, in unidirectional replication an alternative mechanism for priming is needed for leading strand. For ColEl-type plasmids, RNA polymerase performs this role.

Synthesis of leading strand creates a loop of single stranded DNA, which is used as a template for the synthesis of lagging strand. This loop becomes larger, as the unwinding advances (Fig.). The length of this loop should be atleast the size of an Okazaki fragment, before it can be used by the core enzyme for DNA synthesis.

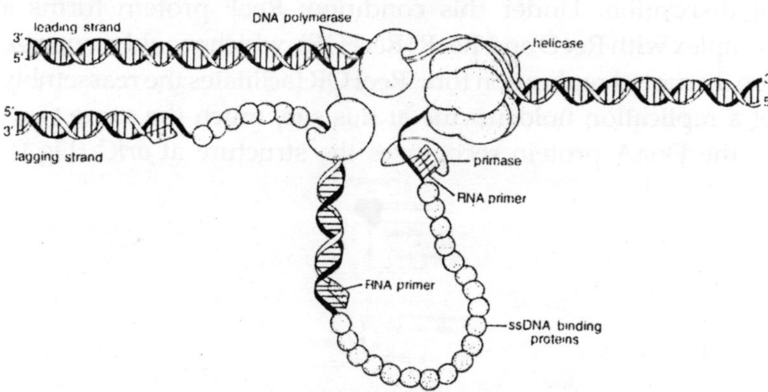

Synthesis of DNA on two strands and associated enlargement of loop due to the synthesis of leading strand.

Discontinuous Synthesis on Lagging Strand : The events on lagging strand are more complex than on leading strand, and DNA polymerase enzyme is less processive. Primase is taken up from solution and activated by DnaB, to synthesize a primer (10-20nt in length) on the lagging strand. The 'DNA pol III HE' interacts with primase and provides the signal for primase to dissociate from the fork. The primers are recognized by the polymerase on the lagging strand and are utilized for synthesis of 2000-4000 Okazaki fragments (each l-2kb in length) at a speed of lkb/second. This requires rapid recycling of 'core' polymerase, so that it finishes synthesis of an Okazaki fragment in one or two seconds and is recycled to the next RNA primer upstream. After completing the synthesis of an Okazaki fragment, core rapidly dissociates, leaving the β_2 ring behind, and rapidly associates with a new β_2 clamp, that is assembled at another primed site upstream. From its new position the core synthesizes yet another Okazaki fragment and the cycle continues (Fig.).

Replication at Disrupted Replication Fork : Sometimes there may be premature disruption of replication, so that the replication machinery (DNA polymerase holoenzyme III) needs to be reassembled to reinitiate DNA replication from the point of disruption. Under this condition, RecF protein forms a complex with RecO and RecR (RecFOR), which would recognize the disrupted replication fork. RecFOR facilitates the reassembly of a replication holoenzyme at this site, much the same way, as the DnaA protein recognizes the structure at *ori*C (Fig.).

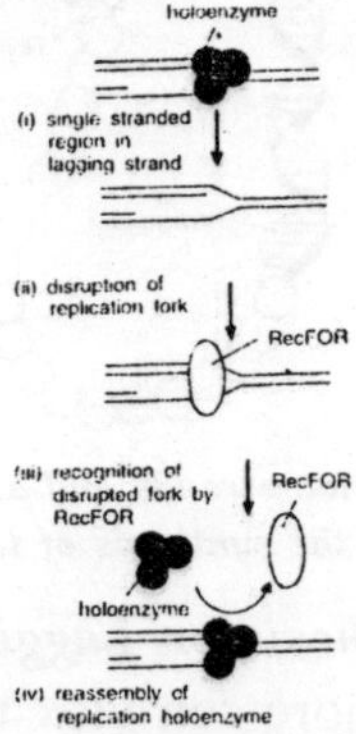

DNA replication at disrupted replication fork, due to reassembly of holoenzyme with the help of RecFOR.

Termination of DNA Replication

The *E. coli* chromosome carries a large termination zone, diametrically opposite from *oriC*, which blocks the progress of replication forks (of bidirectional replication), meeting at this region. *E. coli* chromosome and several plasmids carry specific sequences, called *ter* sites, where TBP (ter binding protein) or 'Tus protein' binds. [In *B. subtilis*, the protein equivalent to Tus is called replication terminator protein (RTP)]. In the termination zone of *E. coli,* there are three ter sites *(terA, terD* and *terE)* for counter-clockwise fork and three *ter* sites *(terB, terC* and *terF)* for clockwise fork. These six sites are arranged in overlapping manner, leaving no 'replication-free' gap on the chromosome (Fig.). *TUS-ter* complexes formed at *'ter'* sites stalls the replication fork, by inhibiting the DNA helicase or DnaB. When this termination zone is deleted, replication stops simply by the meeting of opposite replication forks, suggesting that the termination zone is not essential.

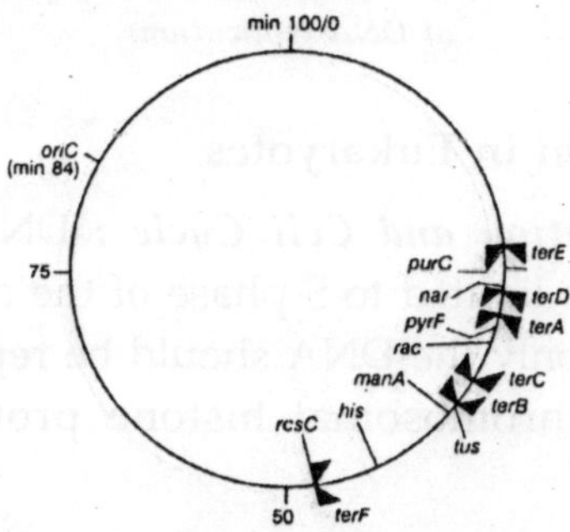

A partial map of E. coli chromosome, showing the position of six termination sites (terA-terF) and the tus gene relative to the location of oriC.

In 1996, X-ray crystal structure of (Tus)-ter complex from *E. coli* was published. The protein-DNA interaction was found to be completely new and explained how the complex arrested the replication fork on its one end (B or blocking end) and not on its other end (P or passage end). The Tus protein in the (Tus)-ter complex is organized into two domains, connected by two antiparallel beta strands, thus creating a central cleft into which DNA fits (Fig.). The Tus protein grasps DNA, and on the right hand provides a barrier (B end of the complex) to the replication fork moving from right to the left. This means that the helicase can not disrupt the tight Tus-ter contacts. The complex, on its other end called P (passage end), does not provide any barrier to the helicase and the replication fork moves through from left to right, so that the complex is displaced and the replication is completed.

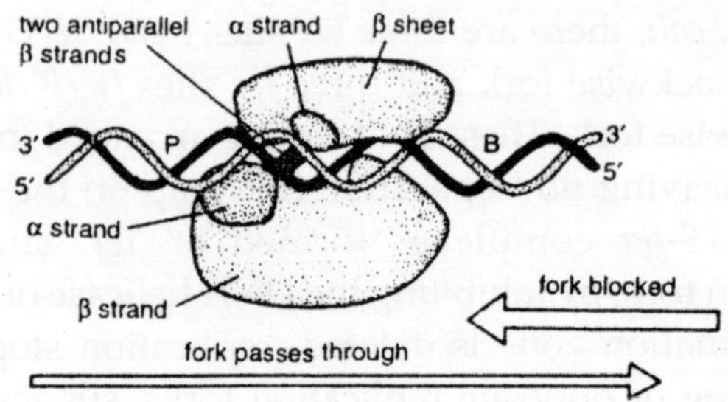

The interaction of Tus protein with ter sites as revealed by the three dimensional structure of (Tus)—ter complex for termination of DNA replication.

DNA Replication in Eukaryotes

DNA Replication and Cell Cycle : DNA replication in eukaryotic cells is limited to S-phase of the cell cycle. During the S-phase, not only the DNA should be replicated, but also the associated chromosomal histone proteins should be duplicated.

The initiation of DNA replication is a prominent landmark in the cell cycle and is, therefore, an important point for exercising control. Further, while in *E. coli* and other prokaryotes, a solitary origin for DNA replication is found, in eukaryotes, DNA replication needs to be initiated at multiple origins (upto several thousands; 10^3 to 10^5 replication events occur in each cell cycle) in a coordinated manner, though not identically at all origins. This leads to great variation in the duration of S-phase (S phase is 3.4 minutes in early embryonic cells, but 10 hours in cultured cells in *Drosophila melanogaster)*, even though the rate of DNA synthesis does not differ.

It is possible that different specific initiator proteins are used at different origins at different times during S-phase. For instance, in yeast, a protein Abflp (a transcription factor) binds many origins, but not all origins of replication. However, a set of origins may be organized in clusters to be used synchronously.

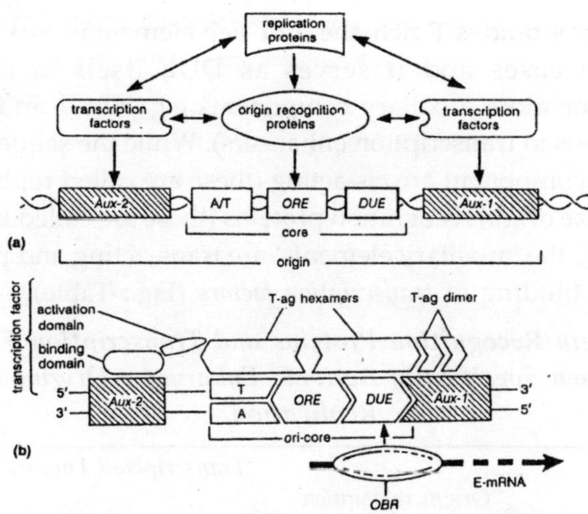

***Details or a eukaryotic origin of DNA replication : (a) different components of* ori; *(b) binding of proteins on different components of* ori.**

Replication Origins and Initiation of DNA Replication (cis and Trans-acting Elements: Replicators and Initiators) : For a study of replication origins, eukaryotes have been classified into (i) simple genomes including those from yeast and a number of eukaryotic viruses and (ii) complex genomes, represented by specific regions of chromosomes belonging to *Xenopus*, Chinese Hamster Ovary (CHO) cell lines, *Drosophila* and other metazoans. While more detailed information is available for simple genomes, the information for complex genomes is both fragmentary and difficult to explain. In simple genomes, an origin of replication encompasses 100-l000bp of. DNA (100-200bp in budding yeast; 500-1000bp in fission yeast), but in complex genomes DNA replication occurs at many sites within an *initiation zone*, although some sites are highly preferred over others.

In simple genomes, *ori* consists of two parts : (i) core component (analogous to transcription promoter); it consists of an *Origin Recognition Element (ORE)*, a *DNA-Unwinding Element (DUE)* and an *A/T-rich element*, (one strand is A-rich and other strand is T-rich; the A/T rich element is part of ORE in some cases and it serves as DUE itself in others); (ii) one or more auxiliary components e.g. *Aux-1* and *Aux-2* (analogous to transcription enhancers). While the sequences of the core component are cis-acting (these are called replicators) and utilize origin recognition proteins (these are called initiator proteins), the 'auxiliary elements' are trans-acting and provide sites for binding of *transcription factors* (Fig.; Table).

Origin Recognition Proteins and Transcription Factors, which Specifically Activate Eukaryotic Origins of Replication.

Origin	*Origin recognition protem(s)*	*Transcription Factors*[a] *Aux-2*	*Aux-1*
SV40 *ori*	T-ag	Spl., T-ag	T-ag
Pyv *ori*	T-ag	AP1	T-ag
BPV *ori*	El	E2	
Ad2ori	[preTP:Ad DNA pol]	NF1 OCT1	
EBV *oriP*	EBNA-1	EBNA-1	
HSV *oriH*	UL9	Several candidates	
mtDNA *oriH*	RNase MRP	mtRNA polymerase	
Yeast ARS	ACBP?, ORC?	ABF1	ABF1

Among simple eukaryotic genomes, replication origins in yeast *(Saccharomyces cerevisiae)* were studied in great detail and were termed 'replicators' or 'autonomously replicating sequences (ARSs)'. It has been shown that these sequences, when present in a plasmid, help high frequency of

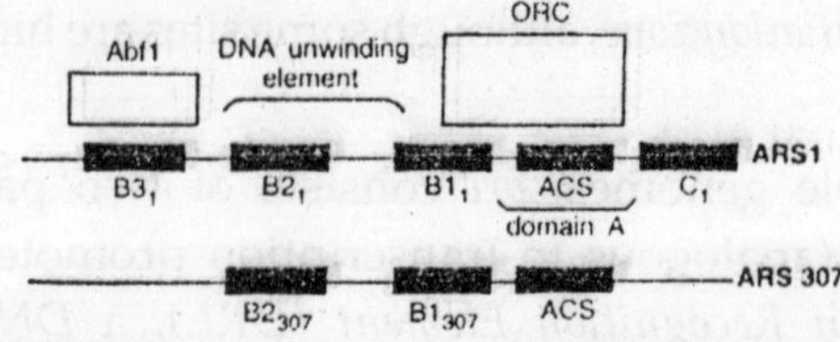

Components of two different auto- nomously replicating sequence elements (ARS1, ARS 307), as DNA replication origins in yeast.

transformation and allow autonomous replication of plasmids in yeast cells. It has been shown that all DNA replication origins in yeast cells are ARSs, although all ARSs are not replication origins. (In all ARSs, meant for DNA replication, there is an llbp *'AT-rich consensus sequence'* or *'ACS'* or *'domain A'*, which is the core component, and is essential but not sufficient for the function of ARS.

The two ARS elements that have been studied in detail are ARS1 and ARS307. Both have domains *A, B* and C. The domain *B* is divided into subdomains *B1, B2* and *B3* in ARS1 and into subdomains *B1* and *B2* in ARS307 (Fig.). Although there is no sequence similarity, but there is functional conservation between these two ARSs (outside ACS). At 3' end of the T-rich strand of ARS, the *'domain B'* is believed to be involved in binding of initiator protein, helix destabilization via DNA unwinding, DNA bending, and binding of accessory protein factors.

The sequences 5' to the T-rich strand of the ACS are similarly referred to as *'domain C.* The domain A of ARS functions in concert with subdomains of domain *B*, each of which is important but not essential for origin's activity. The transcription factor Abflp (ARS binding factorl) binds at B_3 in ARS1 (analogous to auxiliary components) and trans-activates ARS elements, upto lkb away from 'ACS' or 'domain A'.

ORC, MCM, Cdc6p/Cdc18p and Licensing Concept : A search for initiator factors that recognize ARS1 (replicator) in a sequence-specific manner, revealed that a complex of six polypeptides binds ACS element (A) of ARS1 and Bl element with high specificity. This complex is described as 'origin recognition complex (ORC)' and is a eukaryotic 'initiator'. There are also proteins which bind B_3 and B_2. It is believed that ORC functions as a landing pad for many proteins that are important for DNA replication.

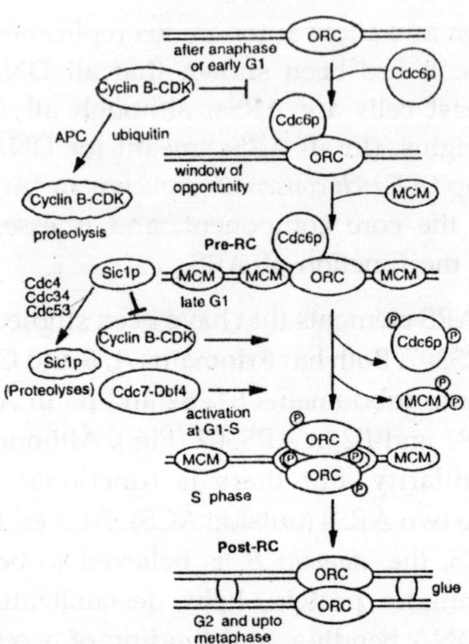

The role of ORC, MCM and Cdc6p/Cdcl8p in granting licence for DNA replication in yeast.

Among these ORC binding proteins, MCM (mini-chromosome or plasmid maintenance) proteins and Cdc6p (in budding yeast) or Cdcl8p (in the fission yeast) are the most important. These MCM proteins are essential for the maintenance of some 'ARS containing plasmids'. It is also speculated that perhaps Cdc6p/Cdcl8p is the loading factor (RLF-B; RLF=replication licensing factor) for the MCM complex at late Gl. Further it is believed that MCM complex, in association with ORC at the origin site must be phosphorylated (by kinases, Cdc7p-Dbf4p, cyclin B-CDK), to initiate replication, thereby removing MCMs and thus granting licence for replication (Fig.). This is described as 'licensing concept.

There are two states of the chromosome during DNA replication: (i) a pre-replication complex (pre-RC) state (with ORC, Cdc6p, MCM, etc), corresponding to competent state of Gl chromosomes and (ii) a post-replication complex (post-RC) state containing only ORC. The pre-RC is activated by cell-cycle regulated activators to cause initiation of DNA replication and is disrupted after the initiation of DNA replication.

The 'origin recognition proteins' listed in Table, help directly in DNA unwinding and indirectly in the initiation of replication through their association with 'replication proteins' including DNA polymerases. The transcription factors, which bind the 'auxiliary elements', have each a 'DNA binding domain' and an 'activation domain', the latter involved in activation of 'origin recognition proteins' through protein-protein interaction (Fig.). The transcription factors thus facilitate DNA replication through the following activities: (i) initiation of transcription through *ori* core to produce RNA primers for DNA synthesis; (ii) the assembly of an initiation complex; (iii) activation of initiation complex and (iv) overcoming repression due to chromatin structure.

In complex genomes, a surprising feature is the appearance of 'replication bubbles' (100-500bp) throughout an 'initiation zone' (5000-55000bp). This suggests that replication starts at many sites within an initiation zone. There is, however, evidence that most replication events in metazoan chromosomes begin at sites, which are not randomly chosen, but are specific sites within the larger initiation zone, which is defined by the presence of replication bubbles. There is also evidence that DNA synthesis occurs by the replication fork mechanism and nucleosome segregation is distributive. The most extensive analysis of mammalian DNA replication (upto 1992) has been carried out in a 300kbp region around the *dhfr* gene of Chinese Hamster Ovary (CHO) cells (dhfr = dihydrofolate reductase). In this 300kbp- region, replication begins in three different initiation zones referred to as *Ori*-α, *Ori*-β, *Ori*-γ. In each of these initiation zones, the origin of replication is described as an 'Origin of Bidirectional Replication (OBR)' associated with several elements, which facilitate replication initiation (Fig.).

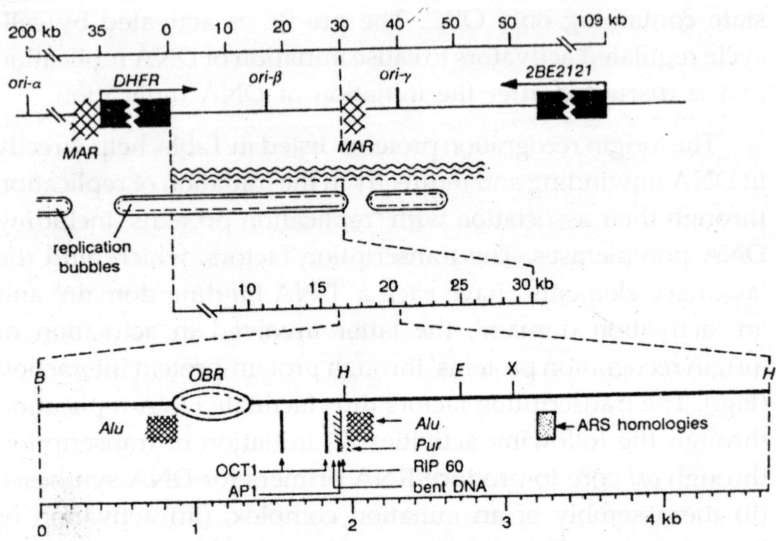

Details of dhfr ***region (in a CHO cell line) with three) initiation zones*** **(ori-a, ori-$, ori-y)** ***lower figure shows details of*** **ori-$** ***(for details, see text).***

The *Ori* β, which is the most extensively studied zone in this region is flanked by two 'matrix attachment regions *(MAR)'.* The *OBR* within *ori*-β is flanked by (i) two *Alu* repeats, (ii) sequences associated with DNA amplification, (iii) autonomously replicating plasmid DNA, (iv) a segment of bent DNA, (v) sequences homologous to yeast ARS consensus sequence, (vi) binding sites for atleast four proteins involved in the initiation of DNA replication (RIP60, PUR, OCT1 and API; the last two are transcription factors).

Although, the mechanism of DNA replication in complex genomes like those of mammals is not fully understood, some generalizations can be made: (i) initiation of DNA in mammalian cells generally occurs in intragenic regions and initiation zones are much longer as indicated above; (ii) the zone where replication is initiated changes during development, this being broad in early embryonic and restricted in later stages (as shown for ribosomal DNA in *Xenopus);* (iii) there are also DNA sequences, called locus controlling region (LCR), which can control positioning of origins of DNA replication to specific sites (as shown for human (β-globin gene).

CDKs and Cyclins in DNA Replication : Cyclin-CDK activity is also needed both for initiation of DNA replication at START point of cell cycle and also for blocking the initiation of DNA replication during and after the S phase of cell cycle. For instance, in budding yeast, B-type/S-phase cyclins (Clb5p, Clb6p) are needed for DNA replication. The activity of Clb5-CDK (*e.g.* Clb5/6-Cdc28) is inhibited by Siclp, which, therefore, needs to be removed to allow initiation of DNA replication. The removal of Siclp is achieved through ubiquitin-dependent proteolysis after START. This proteolysis requires Cdc34p, Cdc53p, Cdc4p and Skplp proteins and also the Gl cyclin-CDKs (Clnlp-Cdc28p, Cln2p-Cdc28p). Thus cyclins and CDKs have been shown to have multiple roles, particularly including the following: (i) activate DNA replication; (ii) prevent re-replication by blocking the formation of pre-RC in G2 and (iii) allow establishment of metaphase. After mitosis, the cycle is reset, by degradation of cyclins due to anaphase promoting complex (APC) and ubiquitin dependent proteolysis.

Cdc7p-Dbf4p for MCM Phosphorylation : The DNA replication in eukaryotes requires another kinase activity (independent of cyclin-CDK) made available through Cdc7p-Dbf4p, which activates competent Gl state. The Cdc7p protein (conserved in yeast to mammals) is a kinase, whose activity depends on Dbf4p (it also interacts with Cdc5p, another kinase) and remains active during Gl to S phase. It is believed that Cdc7p phosphorylates MCM proteins in the pre-RC.

Cln3-Cdc28 Kinase for Transcriptional Control of S Phase Genes : Many genes (~ 30 genes) whose products are needed during S phase, are transcribed during late Gl phase. This late G1-transcription is regulated by Cln3-Cdc28 kinase. The transcription factor MBF, which binds to the MCB elements and the transcription factor SBF, which binds the SCB promoter element, are activated by Cln3-Cdc28 kinase to stimulate expression of some important S-phase genes including B-cyclin genes *(Clb5, Clb6)* and Gl-cyclin genes *(Clnl,Cln2)*. Thus Cdc7p-Dbf4 and Cln3-Cdc28 kinases trigger the initiation of DNA replication.

Comparison of Initiation of DNA Replication with Transcription Initiation : The initiation of DNA replication has also been compared with transcription, due to many similarities :

(i) both involve ordered protein-DNA and protein-protein interactions leading to site identification and positioning of appropriate polymerase at this site;

(ii) transcriptional promoters are heterogeneous regulatory elements and so are replication origins;

(iii) promoters have common core elements, where basic transcription factors bind; replication origins also have core consensus sequences for binding of recognition factors like ORC;

(iv) promoters have flanking sequences besides core elements; accessory factors binding on these flanking sequences, modulate the activity of basic transcription factors and thereby regulate the rate of transcription initiation in response to a variety of stimuli; at replication origins also there are 'accessory control elements' (*e.g.* B_3 at ARS1) or 'auxiliary elements', where trans-acting factors bind and modulate the activity of the replication origin. Since transcription factors are involved both in transcription and replication, they may play an important role in temporal coordination of transcription and replication during the cell cycle (Fig.).

Different Steps Involved in Eukaryotic DNA Replication: During 1987-1989, on the basis of studies on classical enzymology, a proposal was put forward for eukaryotic DNA replication.

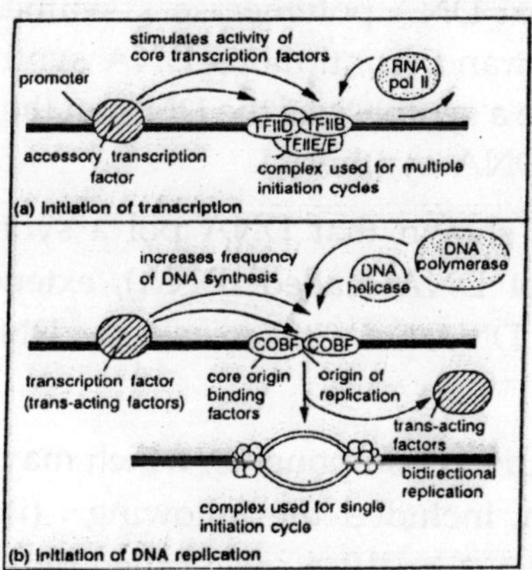

A comparison of initiation of DNA replication with transcroption initiation in eukaryotes.

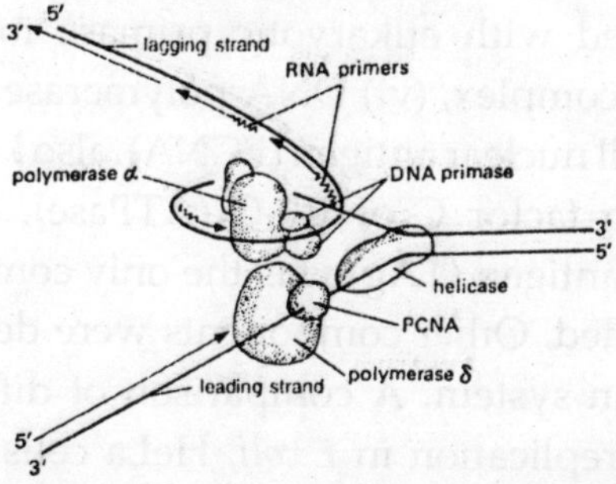

Two different DNA polymerases, synthesizing leading and lagging strands at a eukaryotic DNA replication fork.

According to this proposal, unlike DNA replication in prokaryotes, two different DNA polymerases were required for eukaryotic DNA replication (Fig.). These two DNA polymerases (DNA polymerase a and DNA polymerase δ) were purified and characterized. (Corresponding enzymes in yeast are called DNA polymerase I and DNA polymerase III, which should not be confused with *E.coli* enzymes bearing same designations). In 1990, results of a study were reported where DNA replication (using SV40 genome) in a cell free system depended on T antigen (tumour antigen) derived from SV40 and on a number of purified mammalian proteins. It was demonstrated that DNA polymerase δ synthesizes the DNA on the leading strand (continuous DNA synthesis), whereas DNA polymerase a synthesizes the DNA on the lagging strand (discontinuous DNA synthesis).

Later it was shown that DNA pol a synthesizes only a small segment of DNA (called iDNA), extending the RNA primer and then DNA pol δ takes over the DNA synthesis on both strands).

The eight purified components, which may be involved in DNA replication, included the following : (i) T antigen, (ii) replication protein A or RP-A (also called replication factor A or RFA protein or eukaryotic SSB), (iii) topoisomerase I, (iv) topoisomerase II, (v) DNA polymerase a (pol a), which is tightly associated with eukaryotic primase to form a DNA poly *a*-primase complex, (vi) DNA polymerase δ (pol δ), (vii) proliferating -cell nuclear antigen (PCNA), also known as cyclin, (viii) replication factor C or RF-C (ATPase). Of these eight components, T antigen (TAg) was the only component, which was SV40 encoded. Other components were derived from the host mammalian system. A comparison of different proteins used for DNA replication in *E.coli,* HeLa cells/SV40 and the phage T4 is presented in Table).

TAg has a DNA binding domain and a domain having ATPase and helicase activities. The SV40 *ori,* involved in the initiation of DNA replication, consists of three defined regions extending over 65 bp : (i) an imperfectly inverted repeat sequence, (ii) a 27 bp binding site for TAg (site II) and (iii) a 17bp AT-rich sequence, meant for DNA unwinding. Adjacent to 65 bp sequence, there is another TAg binding site called site I. These features may be shared by many or all eukaryotic DNA replication origins.

Different Proteins Required for Similar Functions at the Replication Forks in Bacterial and Mammalian Systems

	Function	Organisms			
		E. Coli	*Archaea*	*HeLa/SV40*	*Phage T4*
1.	Helicase	DnaB	Dna 2 like, MCM 2 like	T-antigen	g47p
2.	Loading Belicase/primase	DnaC	?	T-antigen	g59p
3.	Single strand maintenance	SSB	RPA-k like	RPA	g32p
4.	Priming	DnaG	Pol α -like	Pol α -/primase	g61p
5.	Sliding clamp	β dinner (β_2)	PCNA-like	PCNA	g54p
6.	Matchmaker or clamp loading (ATPPase)	'γ -complex'	RFC-like	RFC	g44/g62p
7.	Catalysis	Pol III core	Family B DNA	Pol δ	g43p
8.	Holoenzyme dimerization	τ dimer (T_2)	?	?	g43p
9.	RNA removal	Pol I	FEN 1	MFI	g43p
10.	Ligator	Ligase	Ligase	Ligase T	T4 ligase

Following steps are suggested for eukaryotic DNA replication, (i) Before DNA synthesis, there is a presynthesis stage of 8-10 minutes duration, for the formation of unwound DNA complex. This stage requires only three purified proteins, namely T antigen (TAg), RP-A and topoisomerase I or II.

The T-antigen, using its DNA binding domain, forms a multisubunit complex with site I and site II in the presence of ATP and causes local unwinding. More extensive unwinding follows the association of RP-A and a topoisomerase, with the help of DNA helicase component of TAg.

In a number of other viruses also, proteins comparable to TAg are used for the formation of unwound complex for initiation of DNA replication.

(i) The topoisomerases help in unwinding of DNA by altering topology of DNA at the replication fork. RP-A or SSB binds to unwound single stranded DNA.

(ii) The primase, which is tightly associated with DNA polymerase α, undertakes primer RNA synthesis,

(iii) The above priming reaction is unusual; it starts with RNA synthesis, like other priming reactions, but the RNA primer is extended by the DNA polymerase activity of DNA pol α , to give a short DNA sequence (3-4 bases long), called iDNA.

(iv) RF-C binds- to the 3' end of iDNA and loads DNA polymerase δ and PCNA (a cyclin, which works as a processivity factor for DNA pol δ and therefore resembles the sliding clamp β_2 *of E.coli).* RF-C and PCNA help in switching of DNA polymerases, so that pol α is replaced by pol δ , which then continuously synthesizes both the leading and the lagging strands. At one time, it was thought that DNA polymerase α synthesized the lagging strand and that DNA polymerase δ synthesized the leading strand. However, it has been shown now that DNA polymerase a only initiates DNA synthesis, and DNA polymerase δ elongates both the strands (leading and lagging strands).

(v) Another RNA primer-iDNA is then synthesized from the replication fork on lagging strand by pol α primase complex and this step is repeated again and again, till the entire DNA molecule is covered. It is also believed that at each replication fork during the elongation phase of DNA replication, there is only one molecule of DNA polymerase α (involved in the synthesis of iDNA for each Okazaki fragment of the lagging strand) and two molecules of DNA polymerase δ (one synthesizing the leading strand and the other synthesizing the lagging strand).

(vi) The RNA primers are removed using MF1 (a (5'-3') exonuclease) and the gaps are filled by DNA polymerase δ , as by DNA pol I in prokaryotic DNA replication. DNA ligase I seals the sticky ends after gap filling. This scheme of DNA replication involving switching between pol α and pol δ is believed to operate in all eukaryotes (Fig.).

In another study, role of yeast DNA polymerase ε in DNA replication was suggested, so that three DNA polymerases (α, δ and ε) were known to be involved in DNA replication. To accommodate three DNA polymerases functioning at the replication fork, A. Sugino and his colleagues proposed that polymerase α might function only as an initiation polymerase at both leading and lagging strands (only polymerase a has primase activity), whereas polymerase ε and polymerase δ are involved in elongation of the leading and lagging-strands respectively. There are other models, which suggest the role of either the polymerases α and δ or only polymerase a in eukaryotic DNA replication. It is also possible that in different systems, different enzyme systems are involved.

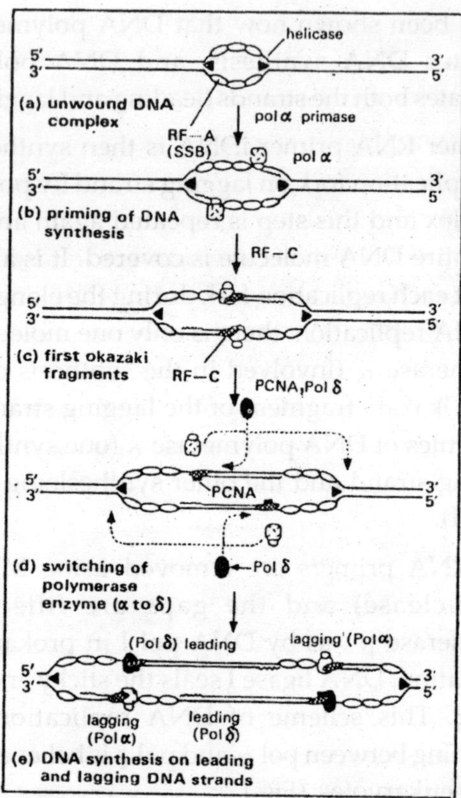

Model for initiating leading and lagging strand synthesis with two DNA polymerases. The first line of the figure shows the unwound complex stage and then four subsequent stages : primer RNA synthesis with primase (priming), first Okazaki fragment synthesis with pol α*, switching of DNA polymerases at the 3′ end of this first Okazaki fragment, and finally, leading and lagging strand synthesis by pol* δ *and pol* α *are indicated. Factors involved are DNA helicase (T antigen in the case for SV40), RF-A, pol a-primase, RF-C, PC.N'A and pol* δ*. The factors are introduced according to the order of entry into the reaction. Wavy lines are, synthesized primer RNAs, and the dashed lines indicate the introduction of pol* δ *and the movement of pol* α*-primase complex along the lagging strand template without dissociation from the replication fork.*

Synthesis of Telomeric DNA by Telomerase : We described a unique feature of the structure of the telomeric ends of eukaryotic chromosomes, characterized by repeated DNA sequences (2-10 bases long). This feature is common to all chromosomes of a species and is conserved. A special ribonucleoprotein molecule (enzymatic in nature) called telomerase uses a special mechanism for the synthesis of DNA at these telomeric ends.

The DNA repeat sequence of telomere has one G-rich strand and the other C rich strand, the G-rich strand having a single stranded overhang. This overhang works as a primer and for its elongation uses as template the RNA component of telomerase enzyme (Fig.). Telomerase synthesizes only the G-rich strand of telomeres. The complementary C-rich strand is perhaps synthesized (extended) by primase-polymerase mediated discontinuous synthesis, typical of semi- conservative DNA replication for which extended G-rich strand is used as a template. For the use of telomerase RNA as template, its sequence should be complementary to telomeric DNA repeat unit, which has been verified. For instance, in *Euplotes,* 5′CAAAACCC CAAAA3' is found in telomerase, which is used for synthesis of 5′GGGGTTTT3' repeats.

The protein component of telomerase has also been studied. It has reverse transcriptase activity to be able to synthesize DNA on RNA template. In 1997, the protein carrying this reverse transcriptase (RT) activity and the gene encoding this protein " were identified in yeast. These proteins were designated p123 and Est2 and the gene(s) encoding these proteins were called *EST* (ever shorter telomeres) genes. These proteins having RT activity *are* also being searched in organisms other than yeast.

Regulation of Telomere Length

In recent years, telomeres have been shown to undergo alteration in their lengths under specific conditions. Following are some examples: (i) telomeres are shorter in tumours; (ii) telomeres shorten, when cells divide in culture; (iii) in certain human tissues, telomeres are shorter in older people than in younger people; (iv) telomeres are longer in sperm cells. Since telomere shortening is associated with senescence and cancer, effort is also being made to establish whether or not telomeres play a causal role in entry to senescence and in causing cancer. Evidence however is available suggesting that reduction in telomere length may not cause aging; instead it may be the consequence of aging.

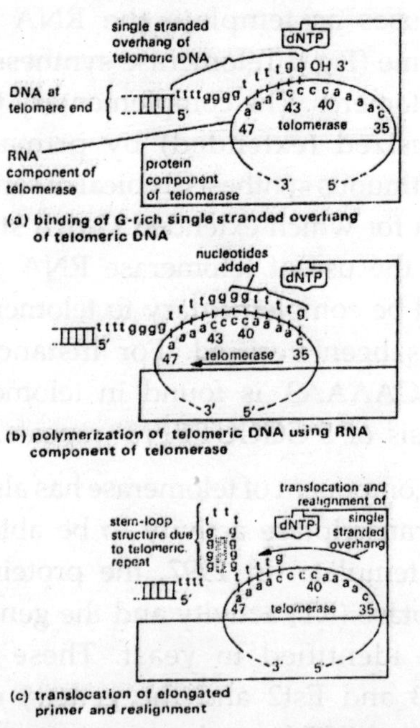

A model for the action of the telomerase from the ciliate Euplotes. This model is based on a model first proposed for the Tetrahymenu telomerase, but in the Euplotes telomerase the template nucleotides have been more accurately defined by in vitro studies, (a) the telomeric primer is bound by interactions with the telomerase through recognition of its G:C base-paired secondary structure and Watson-Crick DNA-RNA base pairing between the 3'-nucleotides of the primer with the telomerase RNA template, (b) polymerization copies out to position 35 on the template, (c) the elongated primer translocates backwards, so its new 3' end, becomes aligned with the template, again by base pairing as shown. The unpairing of the RNA-DNA helix in this step is aided by G:C pairing (indicated by shading) within the newly elongated primer. Another cycle of DNA polymerization occurs, further elongating the primer.

In view of the above, the processes involved in regulating the telomere length have been studied in some detail. Telomerase enzyme specifically elongates telomeres and there are following other factors that influence telomere length: (i) telomere binding proteins; (ii) telomere capping proteins and (iii) DNA replication enzymes.

Models of DNA Replication

Replication Fork Model : The most extensively utilized mechanism of DNA replication involves formation of replication forks, moving in one direction in unidirectional replication and in both directions in bidirectional replication. This method can apply to both linear as well as circular DNA molecules. Replication of circular double helical DNA through formation of replication forks moving in both directions (bidirectional replication) is shown in Figure.

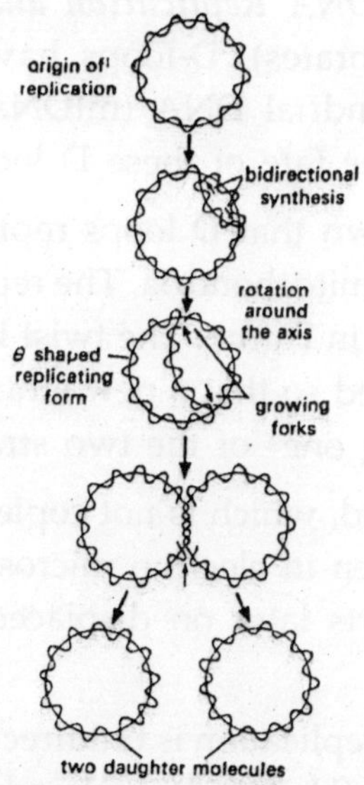

Model for replication of circular double helical DNA, based on formation of double forks due to bidirectional DNA replication.

Rolling Circle Model of DNA Replication : An alternative mechanism has been suggested for the replication of circular

DNA and this is called rolling circle mechanism, which explains multiplication of many viral DNAs and also explains DNA replication during bacterial mating. One of the evidences in support of this model is the observation of long chains of polynucleotides which could measure several times the contour length of circular DNA.

In this mode, due to nicking in one of the two .strands, one strand becomes a polynucleotide with two ends, one end having 3'—OH group, the other having 5'p group. DNA synthesis starts by adding bases at 3'—OH end, displacing the 5' end, which rolls out as a free tail, and is then, copied in $5' \rightarrow 3'$ direction in small segments using fresh enzymes system. These tails thus become double stranded and give rise to a circular DNA later on. A possible mechanism of replication of circular DNA using rolling circle model is shown in Figure.

Mitochondrial DNA Replication and D-loops : In some animal cells (vertebrates), D-loops have been observed in replicating mitochondrial DNA (mtDNA). With the help of radioactive label, the fate of these D loops was studied.

It could be shown that D loops represent early stages of DNA replication in mitochondria. The replication model based on D-loop is shown in Figure. The twist in supercoiled duplex is transiently relieved so that a new strand can be initiated at a fixed point, using one of the two strands.

The other strand, which is not copied, balloons out as D-loop and can be seen in electron microscope at a later stage. The replication starts later on displaced single stranded D-loop.

In this manner replication is unidirectional until about two third of one strand (H-strand) is already synthesized. About 100 minutes are needed for completion of mtDNA replication, and DNA loops in various stages of replication can be seen at a given time.

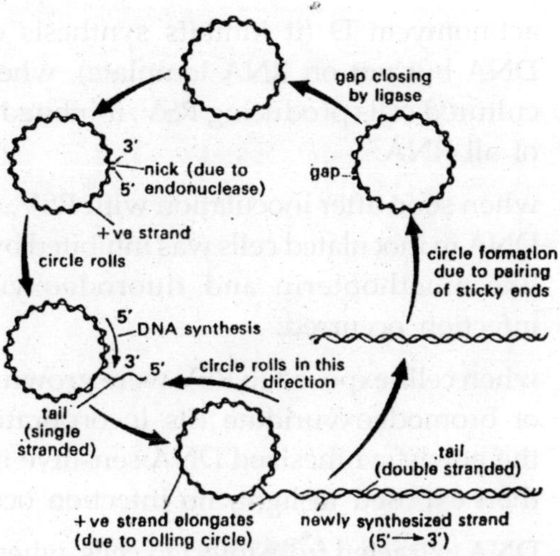

Model for replication of circular double helical DNA, based on rolling of DNA circle following a nick formation due to endonuclease.

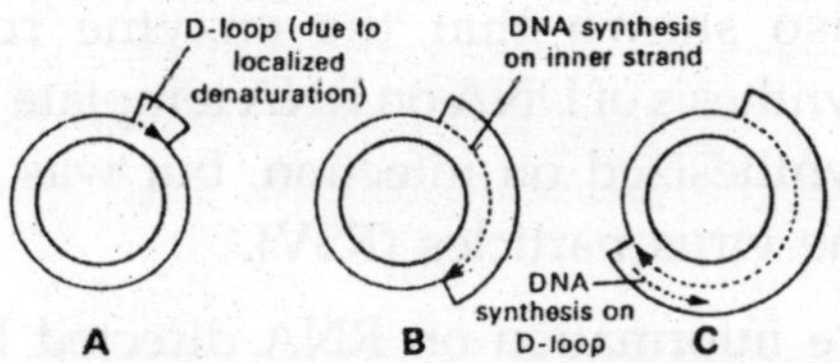

Model for replication of mitochondrial DNA based on D-loop;. D-loop found at A. is extended in B, where replication takes place on inner strand. Replication on displaced strand forming D-loop starts later as shown in C.

RNA Directed DNA Synthesis (Reverse Transcription) : In early 1960's, it was shown that some RNA viruses (e.g. Rous Sarcome Virus = RSV), after infecting the host, give rise to DNA as an intermediate step in their multiplication. This DNA then gives rise to (i) RNA for assembly of viral genomes and (ii) mRNA for synthesis of virus specific proteins. Not all RNA viruses multiply through DNA; rather many of them may actually synthesize RNA directly on RNA as in case of influenza virus or polio virus (Fig.). The evidence for this intermediate synthesis of DNA, often described as' DNA provirus hypothesis, was initially available as follows :

(i) actinomycin D (it inhibits synthesis of RNA on DNA but not on RNA template), when added to cultured cells producing RSV, inhibited production of all RNA;

(ii) when soon after inoculation with RSV, synthesis of DNA in inoculated cells was inhibited by antibiotics like amethopterin and fluorodeoxyuridine, no infection occurred;

(iii) when cells exposed to RSV were grown in presence of bromodeoxyuridine (its incorporation renders the newly synthesized DNA sensitive to light) and then exposed to light, no infection occurred;

(iv) DNA extracted from infected cells, when denatured and hybridized with RNA, double stranded DNA-RNA hybrid molecules could be obtained. It was also shown that the enzyme responsible for synthesis of DNA on RNA template was not freshly synthesized on infection, but was present within the virus particles (RSV).

The above information on RNA directed DNA synthesis was available in 1960's but no enzyme for this activity could be discovered. In 1970, for the first time, disrupted virus particles from RSV (used by H. Temin) and MLV (mouse leukaemia virus; used by D. Baltimore) were shown to undertake DNA synthesis, utilizing dNTPs (one of them made radioactive), suggesting that an enzyme existed in the core of virus particle, which could stimulate RNA-directed DNA synthesis. For this work later both Howard Temin and David Baltimore were awarded Nobel Prize.

The enzyme was initially called 'RNA dependent DNA polymerase', but was later called 'RNA directed DNA polymerase', since the purified enzyme was capable of utilizing a variety of templates including synthetic and natural DNA, RNA and RNA-DNA hybrids. The enzyme subsequently became popularly known as reverse transcriptase. This enzyme resembles other DNA polymerases and was detected in human tumour cells, initially suggesting its role in cancerous growth of cells. However, later this enzyme was found in the normal cells also, thus reducing the initial excitement about this enzyme being the cause of cancer in humans.

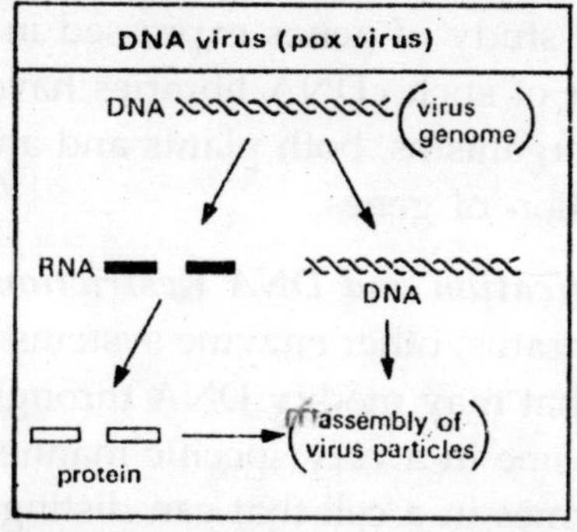

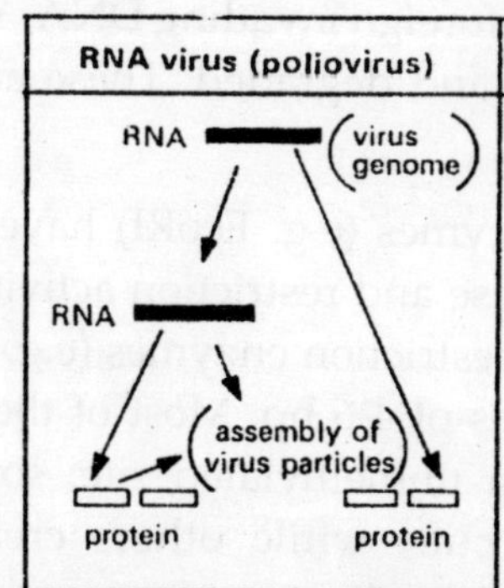

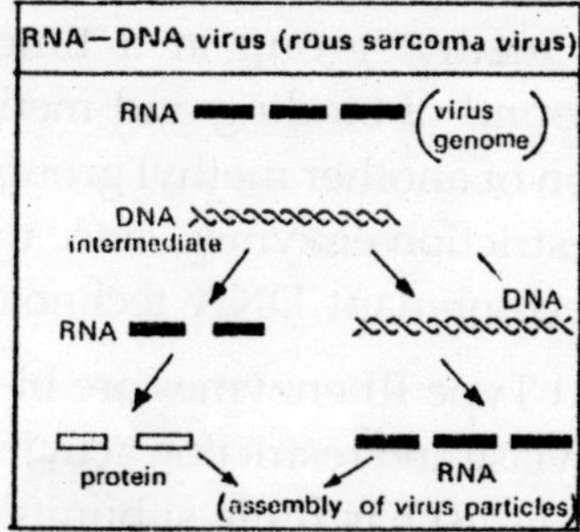

Three kinds of viruses, grouped on the basis of genetic material and mode of replication (a) DNA viruses,(b) RNA viruses and (c) RNA DNA viruses.

Although the discovery of RNA directed DNA polymerase or reverse transcriptase did not help in providing a solution to the problem of human cancer, the enzyme is now being extensively utilized for molecular biology studies. The most important of these uses is the synthesis of complementary DNA (cDNA) from mRNA, leading to the preparation of cDNA libraries for the study of genes expressed in specific tissues. A large number of such cDNA libraries have been prepared in a variety of organisms, both plants and animals, and have facilitated isolation of genes.

DNA Modification and DNA Restriction : Besides DNA replication apparatus, other enzyme systems (methylases) are found in cells that may modify DNA through methylation of adenine or cytosine in a very specific manner. There are also restriction enzymes in a cell that can distinguish between its own DNA and any foreign invading DNA, so that the foreign DNA is attacked and degraded. These enzymes fall in two general classes:

(i) Type II enzymes (e.g. EcoRI) have separate enzymes for methylase and restriction activities; the target sites for type II restriction enzymes (e.g. *EcoRl*, a dimer) are palindromes of 4-6 bp. Most of these enzymes cleave DNA at an unmethylated site; some of them create staggered cuts, while others create blunt ends. A methylase (usually a monomer as in *EcoRl* system) adds one methyl group at a time and dissociates. Another round of binding and methylation is needed for addition of another methyl group. The uses of some of these restriction enzymes, since they are extensively used in recombinant DNA technology,

(ii) Type I and Type III enzymes are bi-functional, having both methylase and restriction activities; a type I enzyme (*e.g. EcoK, EcoB)* has three subunits R, M and S (R for restriction; M for methylation and S for specific target site), but type III *(e.g. EcoP\, EcoP15)* has only two subunits (R for restriction and MS for methylation and recognition). The properties of these three types of enzymes are summarized in Table and a few important restriction enzymes.

DNA Repair

Both in prokaryotic and eukaryotic cells, there are repair enzyme systems to deal with DNA damage, which is caused rather frequently. The changes in DNA leading to damage are broadly divided into two general classes :

(i) Single base changes may be caused by conversion of one base to another (e.g. deamination of 5-methylcytosine to thymine). These are corrected through DNA replication leading to changes in DNA sequences,

(ii) Structural distortions may result from a single strand nick, removal of a base or introduction of a covalent link between bases of same or different strands (*e.g.* formation of thymine dimer due to UV).

A cell may have several systems to deal with DNA damage. These systems include the following :

(i) Direct repair involves reversal of the damage; for instance, photoreactivation of pyrimidine dimer involves reversal of covalent bonds giving the original structure (no fresh DNA synthesis is involved),

(ii) Excision repair involves recognition- of damaged or altered base(s) followed by excision of a sequence including damaged bases. This is achieved by an endonuclease, or an exonuclease. A new stretch of DNA is then synthesized to replace the excised material,

(iii) Mismatch repair involves correction of mismatches or pairing between bases which are not complementary. Mismatches may arise either,

Three Types of DNA Restriction / modification Enzymes

	Property	*Type II enzyme*	*Type III enzyme*	*Type I enzyme*
1.	Protein structure	Separate endonuclease and methylase	Bifunctional; 2 subunits	Bifunctional; 3 subunits
2.	Recognition site	Short sequence (4-8 bp), palindromic	Asymmetric sequence (5-7 bp)	Bipartite; 4 asymmetrical (e.g. $TGAN_8$ TGCT)
3.	Cleavage site	Same as recognition site or close to it	24-26 bp downstrem of recognition site	> 100 bp from recognition site
4.	Restriction & methylation	Separate reactions	Simultaneous	Mutually exclusive
5.	ATP needed or not	No	Yes	Yes

(a) during replication or

(b) due to base conversion (*e.g.* deamination or oxidative damage) and are corrected by a process described as error correction during DNA replication.

Error correction involves a process that distinguishes between 'new' and 'old' strands (due to methylation), so that the error in the newly synthesized strand is corrected.

Direct Repair

As mentioned above, direct repair involves reversal of DNA damage In other words, abnormal chemical bonds between bases or between a nucleotide and an abnormal substituent are broken down, thus restoring the original DNA structure. The enzymes currently known to catalyze this DNA repair include the following :

(i) DNA photolyase repairs cyclobutane pyrimidine dimers (CPD), induced by UV. The enzyme splits the cyclobutane ring, only on activation by light. This enzyme is found in several bacteria (including *E. coli)* and some mammals, but is absent in bacteria like *B. subtilis* and in placental mammals,

(ii) 6-4 photoproduct photolyase deals with the DNA damage involving the formation of '6-4 photoproduct' due to UV. The 6-4 photoproduct is formed firstly due to C4-C6 bond formation between adjacent pyrimidines and secondly, due to migration of a substituent available at C-4 position of one pyrimidine to the C-6 position of adjoining pyrimidine. The enzyme reverses both these anomalies and restores the normal bases,

(iii) Spore photoproduct lyase repairs the lesion caused by UV in *B. subtilis* spores, where a spore photoproduct (5-thyminyl-5,6 dihydrothymine) is produced instead of cyclobutane dimers on UV irradiation. The enzyme breaks the C-C bond between the two thymines in a light-independent reaction,

(iv) O^6-methylguanine DNA methyl transferase (MGMT) is a common enzyme found in all species tested. The enzyme transfers the methyl group of O^6-methylguanine (less efficiently also the other alkyl groups at this position) to a cystene residue on the enzyme. O^6-methylation of DNA is a common effect of several simple alkylating agents including atmospheric methyl halides and products of nitrite metabolism, causing DNA damage.

Excision Repair Systems in E. coli

Excision repair has been classified into the two types on the basis of the nature of excised products, which can either be free bases as in base excision repair (BER), or one or more nucleotides, as in nucleotide excision repair (NER). The lesions involving non-helix distortions (*e.g.* alkylating bases) are repaired by BER, and the lesions that involve distortion of the helical structure of DNA are repaired by NER.

Base Excision Repair (BER) : BER is always initiated by a DNA glycosylase. A number of these glycosylases are known, depending upon the substrate used by each of them. These glycosylases hydrolyse N-glycosyl bonds linking bases to the sugar-phosphate backbone of the DNA, leaving sites that are abasic (without bases). These abasic sites are then acted upon by an apurinic/ apyrimidinic (AP) endonuclease and an AP lyase, so that sugar-phosphate group is removed by hydrolysis of 5'phosphodiester bond. This leaves suitable template-primers for DNA polymerase to synthesize DNA and fill the gap. This results in a repair patch, which can be as small as single nucleotide. Some of the lesions in DNA, that are excised by BER, include uracil, thymine glycols, thymine hydrates, N3-MeAde and 8-oxo-guanine.

Nucleotide Excision Repair (NER) : NER is initiated by endonucleolytic excisions, either only on one side of the lesion (described as endonuclease - exonuclease repair) or on each of the two sides of the lesion (described as excision nuclease or excinuclease repair). In the endonuclease-exonuclease repair, an endonuclease first makes an incision at either the 5' or the 3' end of the lesion. The exonuclease then digests segment of the strand involving the lesion.

The gap is later filled by DNA synthesis and ligation. In excinuclease repair, incisions are made on either side of the lesion by endonuclease activities, giving fragments that are 12-13 nucleotides long in prokaryotes (8th bond on 5' side, and 4th or 5th on 3'side of the lesion) and 24-32 nucleotides long in eukaryotes (20th-25th bond on 5' side and 3rd-8th bond on 3'side of the lesion). It is thus apparent that the distances of the incisions are located at rather precise distances from the damaged base. The enzyme, system is called excision nuclease or excinuclease, and in *E. coli* it is due to combined action of three subunits [UvrA, UvrB, UvrC; also described as (A)BC excinuclease]. After the dual incisions, a gap is formed, to which repair proteins remain associated . These bound repair proteins are dissociated later due to association of replication proteins, which take part in repair synthesis of DNA to fill the gap, followed by ligation.

Although excision repair systems, discussed above, vary in their specificity, the main pathway involves the following three steps: (i) in incision step, an endonuclease cleaves the DNA strand on one or both sides of the damage (7 nucleotides from 5' side and 3 nucleotides from the 3' side); (ii) in excision step, a 5'-3' exonuclease (DNA polymerase I) removes a stretch of the damaged strand (20 nucleotides); (iii) in synthesis step, the single stranded region of DNA is used as a template for synthesis of DNA by DNA polymerase I to replace the excised sequence.

Excision repair, as above, involves different lengths of DNA and is described as (i) very short patch repair (VSP), (ii) short patch repair and (iii) long patch repair. While VSP deals with single base mismatches, the other two deal with more extensive damages, which are repaired through *uvr* genes *(uvrA, B, C, D)*, coding for components of a repair endonuclease. UvrA and UvrB exhibit helicase activity ; UvrD (helicase II) is involved in the disassembly of post-excision complexes.

Mismatch Repair Systems in E. Coli MutHLS Pathway: When there is mismatch in a base pair as in a change from GC to GT, then theoretically it may repair to give rise to either wild type (GC) or to a mutant type (AT). Therefore, the repair system has to distinguish between old and new strands and repair only the new strand to restore the wild type. This requires four proteins, namely MutL, MutS, MutU (UvrD) and MutH coded in *E. coli* respectively by genes *mutL, mutS, mutU (UvrD)* and *mutH*. These *mut* genes are mutator loci, where a mutation leads to increased frequency of spontaneous mutations.

It has been shown that a methylase (dam) coded by *dam* gene, brings about methylation in adenine of sequence GATC on both strands of DNA. Following replication, one strand remains methylated (only at A of GATC sequence) and other remains unmethylated till methylase acts on this new strand to bring about methylation. During this transition period, the unmethylated GATC allows recognition of mismatch by MutL, and binding of MutS to mismatch. MutU helps in unwinding the single strands stabilised by SSB protein and MutH cleaves the newly synthesized strand. Two models for repair have been suggested which are shown in Figure.

In one model MutH cleaves DNA at two GATC sequences flanking the mismatch, while in the other model the cleavage occurs on one side of mismatch at GATC sequence and on the other side at the mismatch itself.

VSP and MutY Pathways : The *mutHLS* pathway of *E.coli* described above corrects mismatches, which arise during DNA replication. There are two other pathways to correct mismaches arising due to action of chemical agents : (a) VSP corrects G-T mismatches (to GC pairs) created by deamination of 5-methylcytosine residues and also corrects G-U mismatches (though less efficiently). The pathway involves participation of the following five proteins (i) Vsr protein nicks DNA strand at a position 5' to the T and is stimulated by (ii) MutS and (iii) MutL; (iv) DNA polymerase I; (v)DNA ligase. (b) The *mutY* gene product corrects mismatches resulting from oxidative damage to G residues, which forms 8-oxo-dGTP, that may be degraded by MutT or removed from DNA by MutM. (8-oxo-G)-A mispair may change G-C to T-A, if not repaired by Mut-Y, with the help of DNA polymerase I and DNA ligase. MutY also corrects A-G and A-C mispairs, though less efficiently.

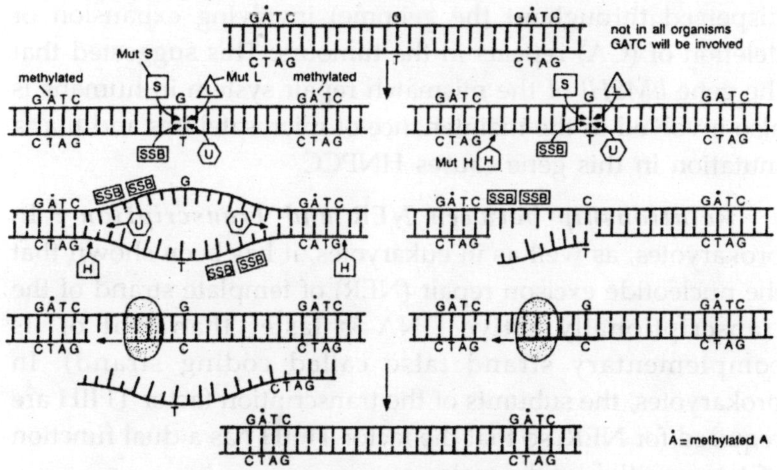

Two models for the mechanism of mismatch repair. In both models proteins called MutL and MutS interact with mismatch site (C—T) and a protein called MutH cleaves the newly synthesized strand. The repair apparatus distinguishes the parental strand from the new one by means of methyl groups (black dots) within the parental GA—TC sequences. The strands surrounding the mismatch are separated with the help of protein MutU and stabilized by SSB.

Mismatch Repair in Eukaryotes (Yeast and Humans) : Eukaryotic mismatch repair genes that are homologous to the corresponding genes in *E. coli,* have been identified in budding yeast *(S. cerevisiae)* and human cells. For instance, in yeast, five PMS genes *(PMS1, PMS2, PMS3, PMS5, PMS6)* have been identified that affect mismatch repair. Of these genes, *PMS1* was shown to be homologous to *mutL* gene of *E. coli.* Six other genes *(MSH1* to *MSH6)* were found to be homologous to *mutS,* of which *MSH2* is the most important for repair of nuclear DNA, while *MSH1* is important for repair of mtDNA.

Atleast four human genes *(hMSH?, hMLHl, hPMSJ, hPMS2)* that have also been shown to be involved in mismatch repair, are linked to hereditary nonpolyposis colon cancer (HNPCC). DNA isolated from colorectal tumours and from HNPCC samples also exhibited microsatellite instability (microsatellites are mono-, di-, tri- and tetra-nucleotide repeats dispersed throughout the genome) involving expansion or deletion of (CA) repeats in the tumours. This suggested that the gene *hMSHl* of the mismatch repair system in humans is also involved in the maintenance of microsatellites and that a mutation in this gene causes HNPCC.

Relationship between NER and Transcription : In prokaryotes, as well as in eukaryotes, it has been shown that the nucleotide excison repair (NER) of template strand of the transcriptionally active DNA is faster than that of its complementary strand (also called coding strand). In prokaryotes, the subunits of the transcription factor TFIIH are required for NER, so that the factor TFIIH has a dual function of transcription and repair.

Recombinational Repair (dimer tolerance) Involves DNA Replication in E. Coli : Dimers, whenever formed (due to UV) in DNA, are reversed either by photoreactivation, due to photolyase (direct repair) or removed by UvrABC catalyzed excision repair. However, some of the dimers still remain and are tolerated by the replication machinery. Although a modified DNA polymerase may replicate DNA, downstream of the dimers, by a mechanism called translesion replication, major mechanisms of dimer tolerance involve the following two recombinational repair activities.-(i) RecBCD- dependent double-stranded break (DSB) repair pathway. This pathway involves extensive DNA replication, which initiates at a D-loop, which is a recombination intermediate. (ii) RecF-dependent post replication gap repair.

In this pathway, RecFOR complex (RecF, RecO, RecR proteins) previously believed to be recombination proteins, were shown in 1997, to assist in replisome reassembly after UV irradiation.

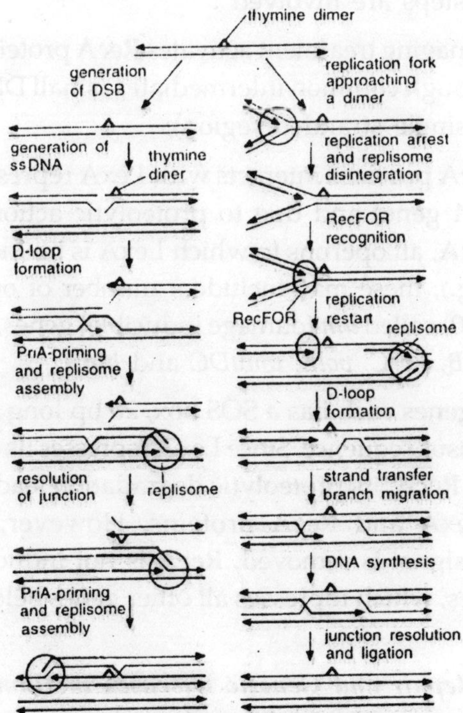

Two different mechanisms of DNA replication for recombinational repair in E. coli.

These two mechanisms of recombinational repair are shown in Figure (also see Fig.). In both the above recombinational repair activities, newly synthesized DNA, after UV irradiation, contains single stranded (ss) gaps opposite the dimers. These gaps are later filled by a mechanism that requires homologous recombination.

An SOS Repair System in E. coli

In *E. coli,* many treatments causing DNA damage or inhibition of DNA replication, induce a complex series of phenotypic changes described as SOS response. The SOS response (increased capacity to repair damaged DNA) is initiated by interaction of RecA protein with LexA repressor. Following steps are involved :

(i) damaging treatment activates RecA protein (a protease), through common intermediates (small DNA molecules or single stranded region);

(ii) RecA protease interacts with LexA repressor (coded by *lexA* gene) and due to proteolytic action of RecA on LexA, all operons to which LexA is bound are induced (Fig.); these may include a number of operons/genes (~30) called *din* (damage inducible) genes, such as *uvrA, uvrB, uvrC, polB, umuDC* and *hinA.*

These genes each has a SOS box, 20 bp long having a 8 bp long consensus sequence. Since LexA represses its own synthesis and that of RecA, its proteolytic degradation leads to synthesis of both LexA and RecA proteins. However, when DNA damaging signal is removed, RecA is not induced and LexA accumulates, which represses all other genes belonging to SOS system.

DNA Repair and Genetic Diseases in Humans : Several genetic diseases in human beings are characterized by cellular hypersensitivity to DNA damage. These diseases also cause increased risk of cancer. Four such recessively inherited diseases include the following : (i) Fanconi anaemia (FA); (ii) Xeroderma pigmentosum (XP); (iii) Ataxia telangiectasia (AT) and (iv) Bloom's syndrome (BS); cells from patients with the first three diseases are uniquely sensitive to DNA damage by DNA cross-linking agents, ultraviolet light and ionizing radiations,those with BS are moderately hypersensitive to several agents causing different types of DNA damage.

Genes responsible for two of these diseases (FA, XP) were identified by complementation of hypersensitive cell lines using DNA Expression clonning technology. The XP DNA products show significant similarity to the products of the RAD DNA repair gene of budding yeast, and ERCC genes of rodents, which are involved in the nucleotide excision repair showing that the essential enzymes involved in DNA metabolism have been conserved through evolution. In future, home work in the molecular mechanism of these and other diseases involving DNA damage will be done.

Eukaryotic Nucleotide-excisipn-repair Genes and Mutants

*Human mutant (and gene)**	*Rodent mutantt*	*Yeast gene*
XP A (XPAC)	?	*RAD14*
XP-B(XPBC)	*ERCC3*	*RAD25 (SSL2)*
XP-C (XPCC)	?	*RAD4?‡*
XP-D (XPDC)	*ERCC2*	*RAD3*
XP-E	?	?
XP-F	?	?
XP-G (XPGC)	*ERCC5*	*RAD2*
CS-A	?	?
CSB (CSBC)	*ERCC6*	?

Contd...

*Human mutant (and gene)**	*Rodent mutantt*	*Yeast gene*
TTD1BR	?	?
?	ERCC1	*RAD10*
?	?	*RADl*
?	?	*RAD7*
?	?	*RADI6*

**XPAC* (and so on) is the term for the gene encoding a protein that corrects XP-A.

‡Rodent complementation groups are numbered. The human genes correcting the defect of rodent mutants are designated *ERCC* (for excision repair cross complementing) genes, the number referring to the number of the corrected group. ‡Level and type of homology not conclusive.

7

Resolution of Gene Structure

Various principles of genetics were explained with the help of a unit of heredity, the gene. Although the term genetics was first used by Bateson in 1905, the term gene (derived from the Greek word genetics, meaning to be born) was proposed in 1909 for the first time by Johannsen. It was subsequently also' regarded as a *unit of function, unit of mutation* and *unit of recombination.* This means that a gene was believed to be a hereditary unit, which was independent in controlling a function leading to appearance of a phenotypic trait for which it was meant. It was also the smallest unit that could undergo a mutational change or could take part in recombination. Such a view was the result of a belief in particulate inheritance, according to which a gene was a locus or a point having no dimensions.

In this chapter, we will discuss the work, which led to changes in our concept of allelomorphs and allowed a study of fine structure of gene at the intragenic level. It will be shown how micromaps for individual genes could be prepared through recombination technique.

Gene vs Allele—A Modified Concept of Allelomorphism

Classical Concept : In classical genetics, a distinction was made between gene and allele on the basis of following two criteria.

Recombination Test : Recombination was believed to take place between two genes but not between two alleles. In other words, intragenic interallelic recombination was not conceived. For instance; a hybrid *aB/Ab* between mutants *aa (aaBB)* and *bb (AAbb)* for two linked genes *A* and *B* could give rise to wild type progeny on a test cross. Since *A* and *B* are linked, this would be possible only due to recombination (Fig.).

On the other hand if two mutants a_1a_1 and a_2a_2 belonging to same gene *A* were crossed and F_1 (a_1/a_2) is test crossed $(a_1a_2 \times aa)$ no wild type progeny would be expected (Fig.). It was thus, earlier believed that different genes or loci could recombine with each other by crossing over but different alleles of a gene could not. This test of allelism is illustrated in Figure.

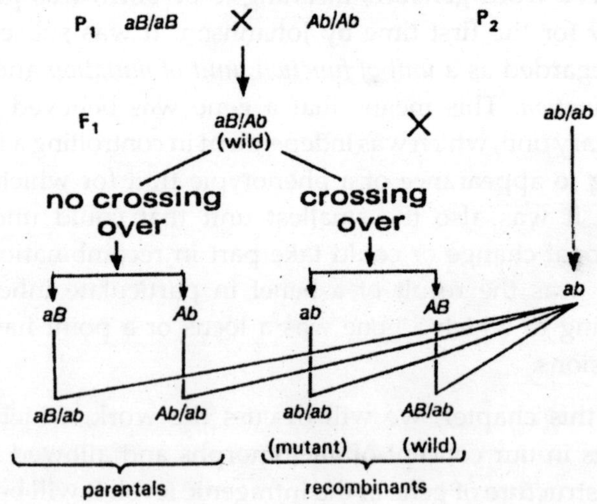

Recombination test (a test cross) in which two linked genes in repulsion phase (aB/Ab) recombine and give rise to wild type individuals.

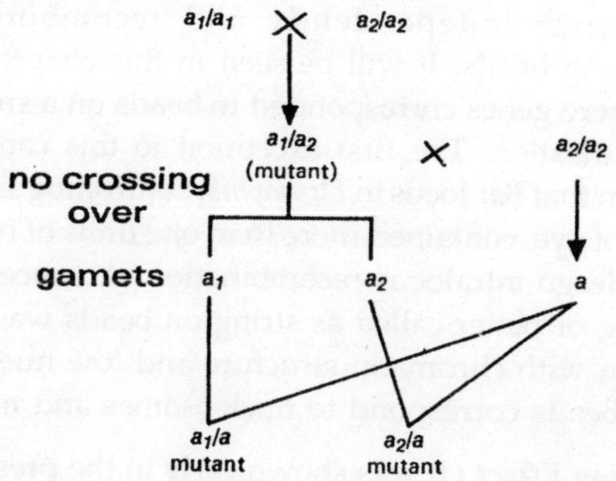

A cross between two mutants of same gene showing lack of recombination as shown by absence of wild type individuals.

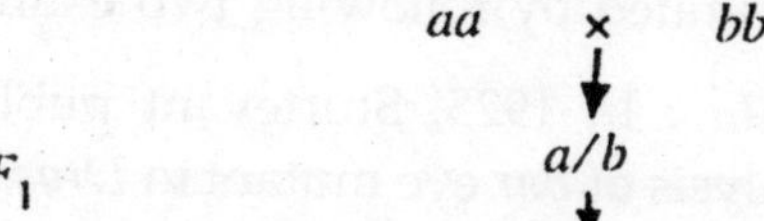

F_2 (1) If wild type appears, a and b are non-allelic.

(2) If no wild type appears, a and b are allelic.

A Test of allelism (classical concept).

Complementation Test : It was also shown that mutant alleles of two different genes coming from two parents, thus being in repulsion phase (also known as *trans* configuration), will complement giving rise to wild type in F_1 generation. But the mutant forms allelic to each other, will never complement (Fig.).

$$\frac{+\,+}{a\;b} \qquad\qquad \frac{a\;+}{+\;b}$$

cis (wild) ***trans (mutant)***

Cis and trans arrangements of alleles a and b in heterozygous condition.

Thus, in its classical concept, the gene recombined as a unit, functioned as a unit and also changed (mutated) as a unit. This concept was consistent with the view of Morgan and his group that genes were like beads on a string, where one bead could change independently and recombine with its neighbouring beads. It will be seen in this chapter that such a view where genes corresponded to beads on a string was an oversimplification. The first exception to this came, when it was shown that *Bar* locus in *Drosophila* controlling size (number of facets) of eye, contained more than one units of function and could undergo intralocus recombination. A concept of beads on a string or better called as string on beads was revived in connection with chromatin structure and 'the nucleosome. In this case beads correspond to nucleosomes and not to genes.

Position Effect : It was shown early in the present century that due to shift of loci from one position to another, or due to shift of another segment in vicinity of a locus, expression changes. This will be illustrated by following two examples.

Bar Eye in Drosophila : In 1925, Sturtevant published results of his study on analysis of *Bar* eye mutant in *Drosophila*. It was shown that it was a semi-dominant mutation leading to reduction in number of facets from 779 in wild type (B^+/B^+)

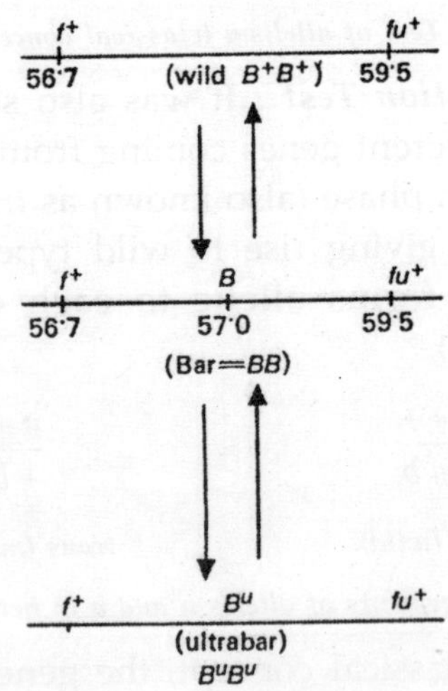

Position effect, due to duplication at bar locus, on the two sides of which genes f^+ (wild allele for forked bristles) and fu^+ (wild allele for fused veins) are shown.

to 358 in heterozygote *(B^+/B)* and to 68 in homozygote dominant *(B/B)*. From homozygous for Bar eye *(B/B)*, mutations to wild type occurred at a frequency of one in 1600, and mutation to an extreme type known as *ultrabar* (with only 24 facets) also occurred, although with a lower frequency (Fig.). Pure *ultrabar* cultures *(B^u/B^u)* could also be established. *Ultrabar* also gave rise to wild type eye with the same rate with which *ultrabar* was obtained from them.

It was observed by Sturtevant that frequency of these events was much higher than what was expected due to spontaneous mutations. Therefore, Sturtevant had to seek an alternative explanation. He formulated a hypothesis and confirmed that new types arose due to unequal crossing over in the region of *Bar* locus. The technique used for confirming unequal crossing over, involved use of marker genes on either side of *Bar* locus, so that if crossing over occurred, then these marker genes on either side of *Bar* locus would also recombine, and will prove the occurrence of recombination. As shown in Figure, asymmetric pairing of homologous chromosomes may lead to unequal crossing over, which is evident from recombined marker genes *(forked bristles = f; fused veins = fu)*. For instance, *fB+/+Bfu*, when test crossed with *fBfu/fBfu*, could give rise to four types

(1) ultrabar, forked .bristles and fused veins *(fBBfu)*

(2) wild type for eye, bristles and veins (+ B^+ +)

(3) wild type eye, forked bristles and fused veins *(fB^+fu)*

(4) ultrabar, but wild type bristles and veins (+ *BB* +)

Sturtevant also demonstrated that phenotype of flies with two bar genes on each of the two chromosomes (double bar homozygous), was different from those with one bar gene on one chromosome and three on the other (Figs.). This indicates that position of gene with respect to adjacent regions also influences its expression. This was called position effect. Later

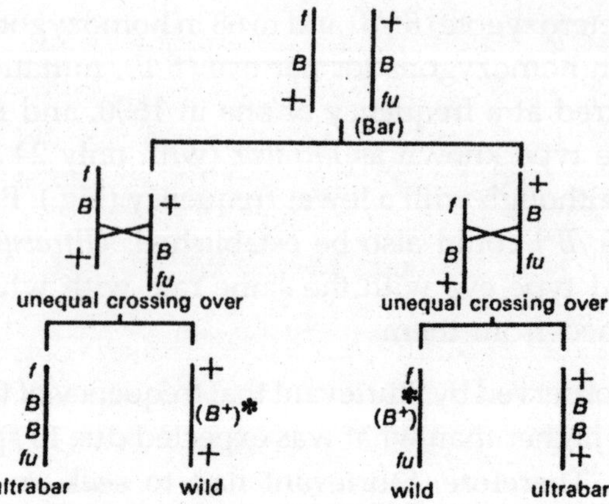

Appearance of ultrabar and wild from bar eyed Drosophila, due to unequal crossing over. The evidence for crossing over was available due to recombination in f (forked bristles) and fu (fused veins) genes on the two sides of bar locus (B). (B⁺)* mean, no bar gene (B) is present or that only one dose of 16A region is found.

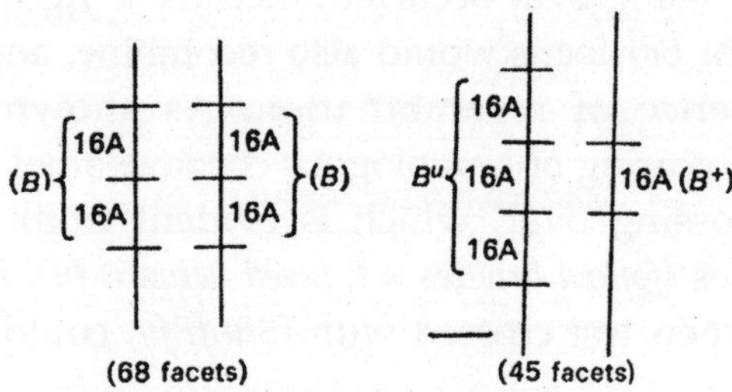

The difference in phenotypes, between homozygous bar and heterozygous ultrabar, although in each case number of I6A segments is same. This illustrates position effect.

C.B. Bridges and his group, through a study of salivary gland chromosomes demonstrated cytologically that *Bar* character was associated with a repeat of 16A region in X-chromosome (Fig.) and that ultrabar (double bar) had two such repeats in the same region.

The following conclusions can be made from this work : *first* that gene is not a point, but has its dimensions and that its alleles may differ in size; *second*, that *alleles* of a gene or locus may recombine with each other, producing new combinations and *third,* that phenotypes of a heterozygote in *cis* configuration (for instance + +/a_1a_2) and that of a heterozygote in *trans* configuration (a_1 +/+ a_2) may differ due to position effect.

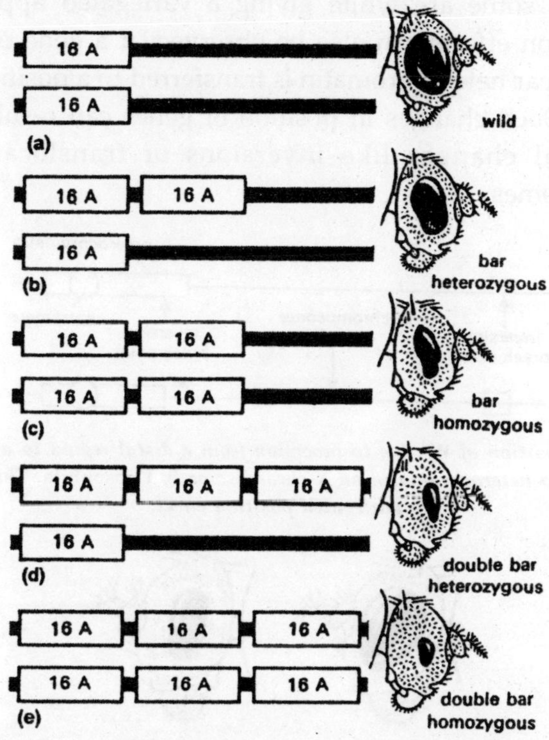

Different arrangements of segment 16A on X-chromosomes in female Drosophila and resulting phenotypes, showing position effect.

Variegated Position Effect : Another kind of position effect can be illustrated with the help of an example of eye colour in *Drosophila.* White eye locus is present on X chromosome near its tip. Further, centromere is flanked on either side with heterochromatin (inert and more condensed chromatin material relative to euchromatin, which is active and less condensed), the remaining region being largely *euchromatic,* so that white eye locus lies in *euchromatin* and quite away from *heterochromatin.* It was shown that when wild type allele of white eye locus *i.e.* w^+, responsible for red eye colour, is transposed to a region near heterochromatin, a mottling of eye is observed (Fig.). Mottling means that some facets are wild type and some are white giving a variegated appearance. Variegation effects can also be observed, if a gene originally located near heterochromatin is transferred to a position away from it. Such changes in position of genes can result due to structural changes like inversions or translocations in chromosomes.

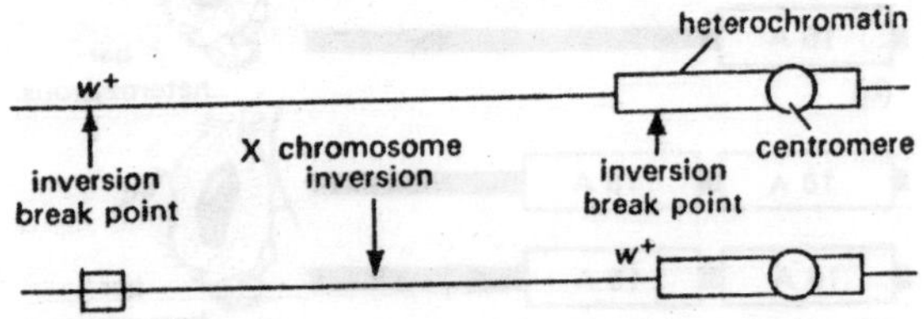

Transposition of W⁺ due to inversion from a distal region to a region proximal to heterochromatin on X-chromosome in* Drosophila. *This leads to variegated position effect.

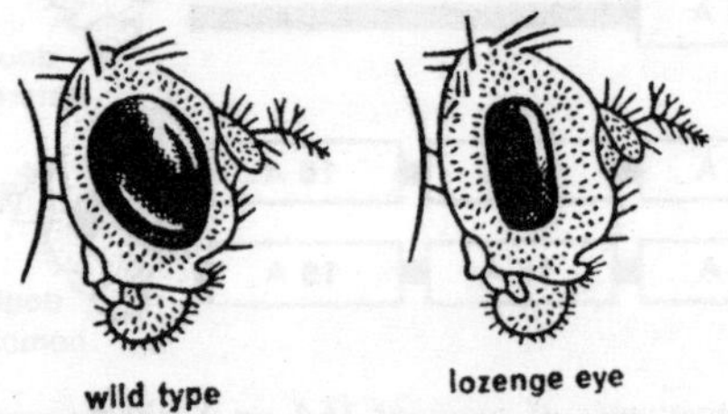

***Wild and lozenge eye phenotypes in* Drosophila.**

Pseudoalleles and Complex Loci : Earlier in this chapter, we discussed criteria for allelism, which included lack of recombination and lack of complementation between two mutant alleles of the same gene. This classical concept of allelomorphism needed change, in view of intensive work done after 1940, initially in *Drosophila* and later in several other organisms.

In *Drosophila,* a recessive sex linked mutant called lozenge *(lz)* is responsible for smaller, darker and more elliptically shaped eyes (Fig.). A number of mutants at this locus were discovered and could be identified due to severity of effect on eye and also due to other phenotypic effects. Ordinarily a fly heterozygous for two *lozenge* mutant alleles (lz^1/lz^2) will not yield wild type in its progeny, because wild allele is absent. However in 1940, Oliver discovered wild type flies in progeny of such heterozygotes, with a frequency of one in several thousands. This frequency was still much higher than expected on the basis of known rates of spontaneous mutations. In view of this, the results could not be easily explained and the lozenge locus was then thoroughly studied by M.M. Green, a student of Oliver.

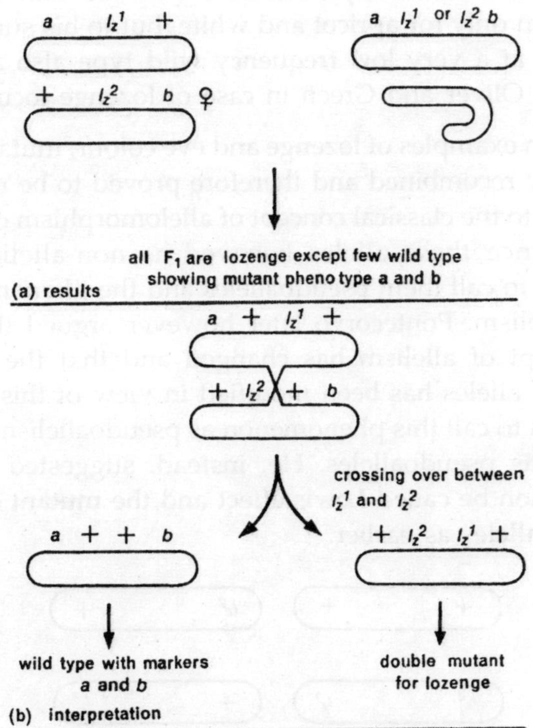

Results and interpretation of Green's experiment in Drosophila where female lz^1/lz^2 was crossed with male lz^1/— or Iz^2/—.

Following Sturtevant's technique with *Bar* locus, Green marked the *lozenge* locus on either side by marker genes to confirm if the wild type appeared due to recombination. If wild type was the result of recombination, Green expected that marker genes would also recombine. In his experiments, Green observed that not only wild type flies appeared for lozenge, but the marker genes *a* and *b* also recombined suggesting that lz^1 and lz^2 could recombine (Fig.). This suggested that two mutants may be separated by a distance within the gene.

Later E.B. Lewis, in 1951 reported results of one of his experiments, where from a cross of *apricot* eyed and *white* eyed flies (apricot is an allele of white eye locus), he obtained F_1 having intermediate eye colour. In F_2, he had expected segregation only for apricot and white, but to his surprise, he recovered at a very low frequency wild type also as earlier shown by Oliver and Green in case of lozenge locus.

In both examples of lozenge and eye colour, mutant alleles apparently recombined and therefore proved to be nonallelic according to the classical concept of allelomorphism described earlier. Since these alleles behaved as non-allelic, Lewis preferred to call them pseudoalleles and the phenomenon as pseudoallelism. Pontecorvo later however argued that since our concept of allelism has changed and that the classical concept of alleles has been modified in view of this work, it is needless to call this phenomenon as pseudoallelism and the elements as pseudoalleles. He, instead, suggested that the phenomenon be called Lewis effect and the mutant elements be called alleles as earlier.

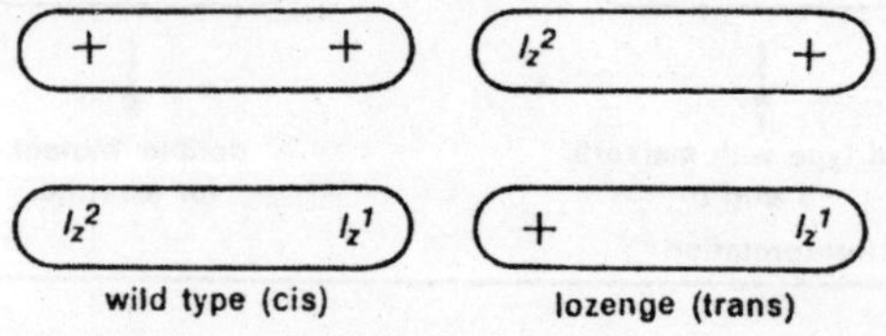

Cis and trans configurations of heterozygotes involving two alleles Iz^1 and Iz^2 at lozenge eye locus in Drosophila.

Position Pseudoallelism or cis-trans Effect : In above experiments of Green and Lewis we tried to demonstrate that intragenic interallelic recombination can take place leading to appearance of wild type (in a very small frequency) in the progeny of a F_1 heterozygote for two mutant alleles. If two mutant alleles occupy different positions, showing close linkage rather than same position, then one may expect that F, heterozygote should have a genotype lz^1 +/+lz^2 or *apr*+/+*w* and, therefore, should express wild phenotype. Since F_1 did not exhibit wild phenotype, we may like to examine the second criterion for allelism *i.e.* lack of complementation. In view of this criterion, we may explain the absence of wild type in F_1 due to lack of complementation. Such a lack of complementation is due to position effect, which can be understood by studying two arrangements of mutant alleles.

For instance in heterozygote for Iz^1 and Iz^2, two arrangements are possible (Fig.), (i) one with both mutant alleles on same chromosme and their wild counterparts on the other homologue (cis), (ii) the other with two mutant alleles on two different homologues (trans). For lozenge locus (Iz^1 and Iz^2) cis and trans arrangements are shown in Figure. It is also indicated that in *cis* position they will express wild type, and in *trans* configuration, the mutant phenotype. This may presumably be due to position effect, where a mutant allele does not allow adjacent region to express wild phenotype. This difference in phenotypes due to *cis* and *trans* arrangements is known as *cis-trans* effect.

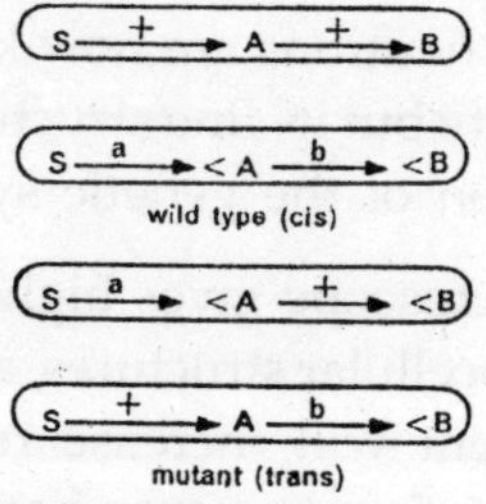

Possible explanation at the molecular level in terms of synthesis of polypeptides in cis and trans configurations.

An explanation for cis-trans effect was given by E.B. Lewis, who believed in sequential nature of gene product synthesis. According to him, if both mutant alleles are on one chromosome, the other chromosome will be normal and will be able to produce the end product. Such a situation is found in cis arrangement. But in trans arrangement, sequence of steps involved in synthesis will be interrupted due to mutations on either of the two homologous chromosomes, thus leading to a mutant phenotype (Fig.).

Fine Structure of Gene

(Using recombinaiion among mutants)

Above explanation makes it clear that a gene expresses itself through a series of steps involved in a sequential synthesis of a product and, therefore, may have one or more functional units. Earlier, we have also shown that there can be several sites in a gene, each capable of being independently involved in mutational and recombinational events. A gene thus is neither a functional, nor a mutational or a recombinational unit, but is a complex locus, whose fine structure should be studied.

Such fine structure has been studied in a number of cases using higher resolving power of recombination technique. One may like to consider analogies, where a cell is a unit of life, but higher resolution through sophisticated microscopes discovered subcellular particles, and also where an atom is a unit of structure of an element, but subatomic particles like electrons, protons and neutrons are now known. By corollary, gene is a unit of heredity but its fine structure has been studied through high resolution of the genetic systems.

As an electron microscope gives higher resolution for the study of cell and its subcellular structures, a very big population in a genetic experiment will increase resolution of genetic system. Fine structure of a gene was not discovered early in the present century, because limited populations were used for recombination studies, which was unavoidable due to the nature of organisms used for study. When bigger populations were studied, intragenic recombination, whose frequency would be very low, could be discovered and a gene could be resolved into smaller units. This will be illustrated with the help of few examples.

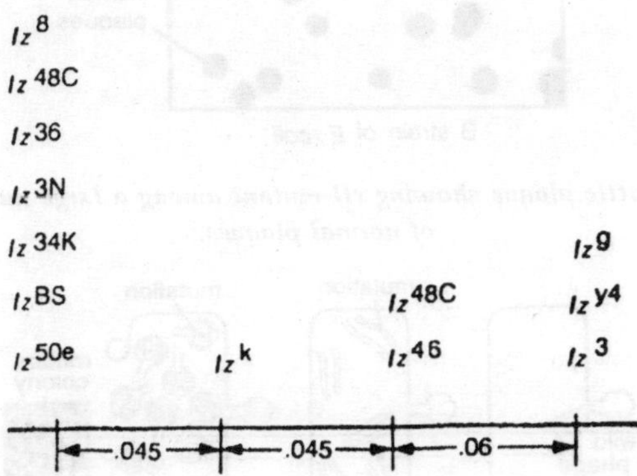

A number of alleles of lozenge occupying four different mutational sites. Recombination frequencies between sites are also given, (redrawn from Strickberger : Genetics).

Fine Structure of Lozenge Locus in Drosophila : The results of Green and Green (1949) and Green (1961) on intragenic recombination in lozenge locus in *Drosophila* demonstrated that 14 alleles can be located on atleast four mutational sites separated by distances proportionate to recombination frequencies. These results are presented in Figure. The alleles on different mutational sites could be subjected to cis-trans test, and complementation relationships showed that all 14 alleles exhibit lack of complementation among each other, so that they belong to only one functional unit.

Fine Structure of rII locus in T_4 Phage : Distinction between wild type and *rII* mutants using K strain of *E. coli.* The most refined analysis of a single gene ever conducted is the one undertaken by Seymour Benzer for a locus in T_4 bacteriophage infecting *E. coli.* This locus is known as *rII* locus and a mutant

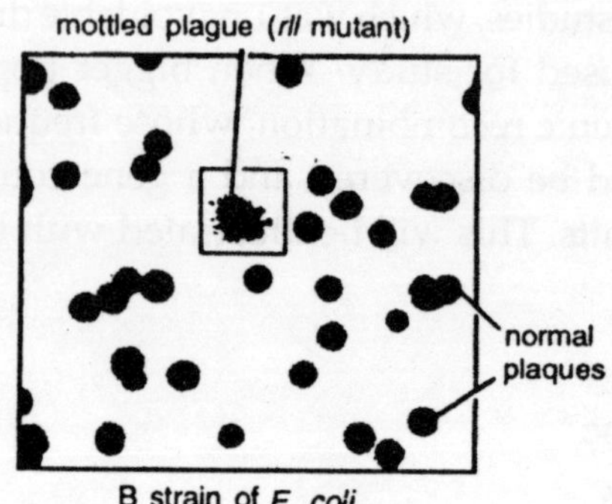

A mottle plaque showing rII mutant among a large number of normal plaques.

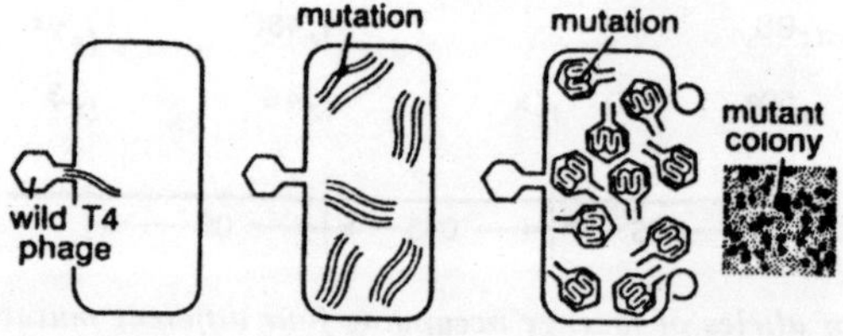

Production of a rII mutant of T_4 phage after infection of E. coli by wild T_4 phage, (redrawn from Set. Amer., Jan., 1962).

phage genotype	strains of *E. coli*	
	B	K
rII+	small fuzzy plaques	small fuzzy plaques
rII	large sharp plaques	no plaques

Expression of wild and rII mutant T4 phages on two strains (B and K) of E. coli.

at this locus is responsible for the formation of rough plaques or colonies (Figs.). This locus had largest number of rapid lysing *(r)* mutants, and is called *rll* locus. It can be distinguished from other *r* loci, by the inability of *rll* mutants to produce plaques on lysogenic 'K' strain of E *coli* (Figs.), which carries

A, prophage, for lysogeny and prophage). The *rll* mutants, may though infect 'K' strain, but can not cause lysis and are, therefore, unable to produce any plaques (plaque is a clear space on petri dish formed due to destruction of bacterial cells due to infection by phage and is characteristic of a phage). In contrast, these *rll* mutants make large sharp plaques on *E. coli,*

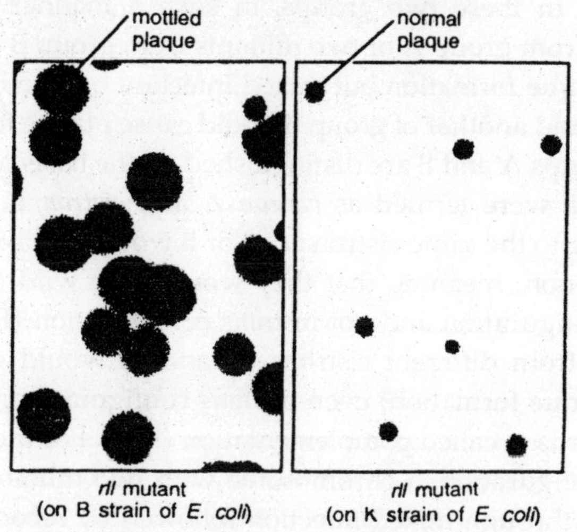

Plaques formed by rII mutant strain on B and K strains of E. coli (normal plaques on K strain are due to wild type T_4).

strain B. The wild type phage T_4 (*rII*$^+$) will make *small* and *fuzzy plaques,* both on B and K strains (Fig.). Further, when 'K' was infected simultaneously by *rII*$^+$ and *rII,* large plaques were formed, since *rII*$^+$ helps in lysis so that *rII* may express. These distinguishing features enabled Benzer to distinguish mutants and wild type phages with high efficiency.

Complementation Test : In order to find out complementation relations between different *rll* mutant alleles, Benzer used two different *rll* mutants, arbitrarily designated as *rII*^{x} and *rII*^{y}. He allowed mixed infection of K strain by these two mutants. Although in most cases, this does not result into lysis and plaque formation, in some cases it does lead to plaque formation. If two mutants did not form plaques on mixed infection, they were placed in the same group, but if plaques are produced, the two mutants involved in mixed infection were placed in two different groups. In this manner, two groups A and B could be established in *rII* region (Fig.). All mutants with the help of complementation test could be classified in these two groups, in such a manner that two mutants from group A or two mutants from group B could not cause plaque formation but mixed infection by one mutant of group A and another of group B, could cause plaque formation. Since groups A and B are distinguished on the basis of *cis-trans* test, these were termed as *cistron A* and *cistron* B. Mutants belonging to the same cistron *i.e.* A or B would exhibit *cis-trans* phenomenon, meaning that they would give wild type only in *cis* configuration and not in *trans* configuration (Fig.). Two mutants from different cistrons (A and B) would give wild type (plaque formation) even in *trans* configuration, which in other words is called complementation (Fig.). For the purpose of *cis* configuration, a chromosome with two mutants can be obtained through mixed infection followed by recombination (Fig.).

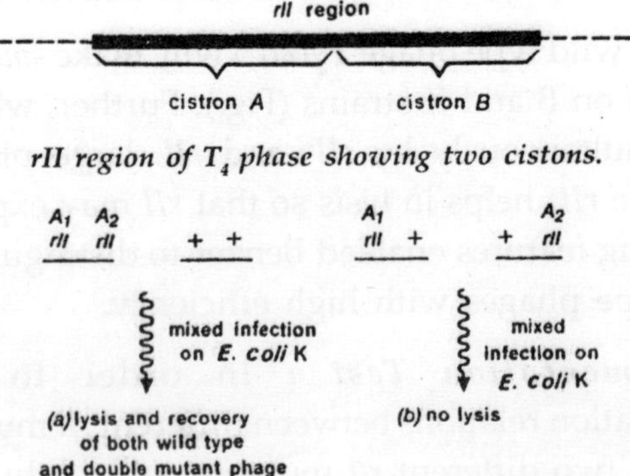

rII region of T_4 phase showing two cistons.

Results of mixed infection by (a) a double mutant (having two rII mutations A_1 and A_2 belonging to same cistron A) and the wild type strains T_4 phage; (b) two T_4 phage mutant strains, having different rII mutations (A_1 and A_2) belonging to same cistron A.

From the complementation test, it is obvious that in *rII* region, two cistrons A and B are independent functionally and must be responsible for sequential synthesis of two separate products, which presumably are polypeptide chains. Therefore,

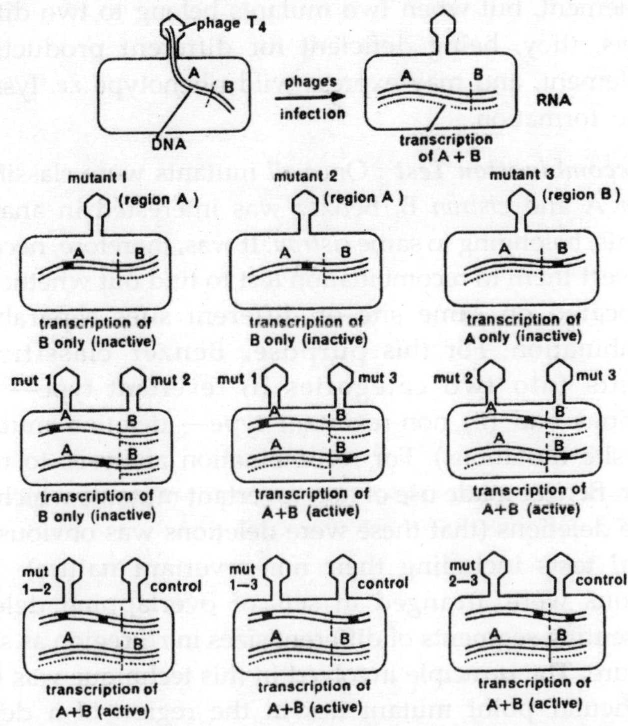

Complementation test between mutants to find out if they belong to same or different cistrons. Last row shows cis test to confirm the same relationship, using double mutants (redrawn from Sci. Amer., Jan., 1962).

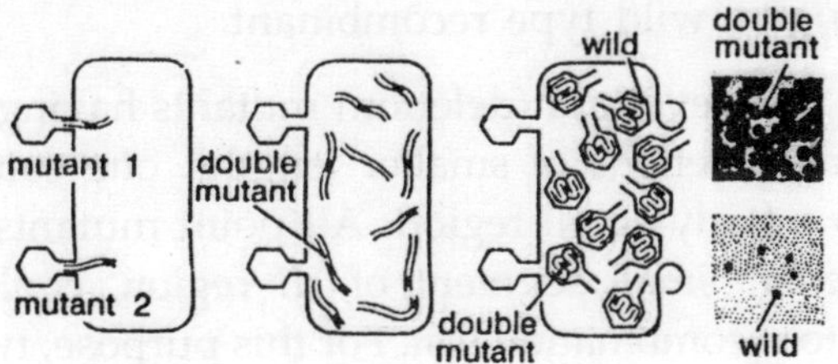

Production of double mutant rII locus due to simultaneous infection of E. coli by two different mutants (redrawn from Set Amer., Jan., 1962).

all mutants belonging to one cistron share a common deficiency, which is different from the deficiency due to mutants belonging to the second cistron. When two mutants belong to same cistron, both are deficient for same product and therefore, they can not complement, but when two mutants belong to two different *cistrons,* they, being deficient for different products, can complement, and may express wild phenotype *i.e.* lysis and plaque formation.

Recombination Test : Once *rll* mutants were classified in *cistron* A and *cistron* B, Benzer was interested in analysing mutants, belonging to same *cistron.* It was, therefore, necessary to subject them to recombination test to find out whether they are located on same site or different sites separable by recombination. For this purpose, Benzer classified all mutants into two categories (i) revertant type— point mutations and (ii) non-revertant type— deletion mutations (multisite mutations). For recombination analysis, to reduce labour, Benzer made use of non-revertant mutants which were due to deletions (that these were deletions was obvious from several tests including their non-revertant nature). These deletions were arranged in sets of overlapping deletions representing segments of different sizes in *rII* region as shown in Figure. The principle involved in this technique was that if a particular point mutant lies in the region of a deletion represented by a *rII* mutant, then on mixed infection with this deletion mutant, the point mutant will not be able to give rise to a wild type, but if it falls outside the deletion region, it will be able to give wild type recombinant.

Using non-revertant (deletion) mutants having successive overlapping deletions of smaller lengths, one could locate a mutant to a fairly small region. All point mutants located in this particular small segment of *rII* region, could, then be subjected to recombination test. For this purpose, two mutants at a time were used for mixed infection on *E. coli* B strain and the lysate produced on plaques formed on B strain was used for infection on K strain to find out the frequency of wild type phage particles produced, because this will represent the recombination frequency. By this method, Benzer was successful in estimating recombination frequencies as low as .0001 per cent, although later Tessman (1965), while studying ultrafine structure of *rII* region recorded frequencies as low as .00001 per cent. Thus Benzer was not only able to divide *rII* region into *cistrons* A and B, but was able to classify mutants belonging to same *cistron* into hundreds of mutational sites separable due to recombination.

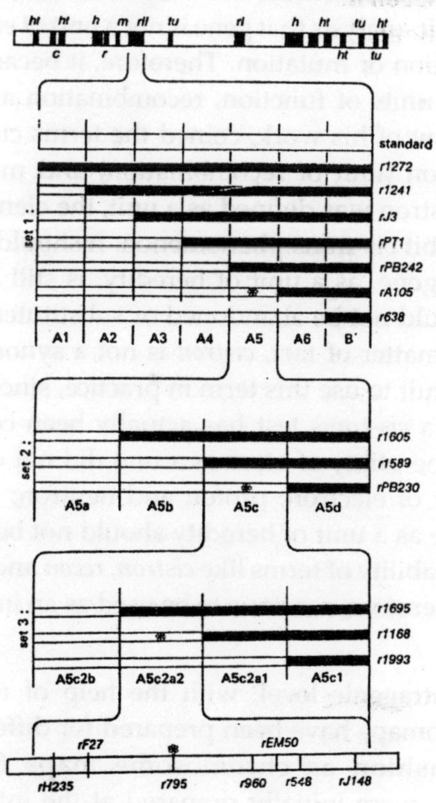

Different overlapping non-revertant (deletion) mutants at rII locus in T_4 phage and their use in locating the position of a revertant (point) mutation (marked as *). Note that three sets of overlapping deletion mutations cover narrowing range of areas of rII locus (rII, A5, A5c, A5c2a2) were used.

Through recombination analysis, about 2000 mutations were mapped in *rII* region and were found by Benzer to occupy 308 different sites called mutational sites. These sites were, although scattered fairly evenly throughout the length, some of them were called hot spots since they had numerous mutations relative to other sites. In contrast, some sites had no mutants and were, therefore, called zero sites. Benzer eventually estimated 400-500 mutational sites in *rII* region and called each of them a unit of mutation or a *muton.*

Cistron, Recon and Muton : As we have seen, it was proved through genetic analysis that gene is not a unit of either function or recombination or mutation. Therefore, it became necessary to define the units of function, recombination and mutation. Benzer, in view of his work, coined the terms cistron (unit of function), recon (unit of recombination) and muton (unit of mutation). Cistron was defined as a unit, the elements (alleles) of which exhibit cis-trans phenomenon. It should however, be realized that gene, as a unit of heredity, is still a very useful term and should not be abandoned or substituted by the term *cistron.* As a matter of fact, *cistron* is not a synonym for gene and it is difficult to use this term in practice, since you can not use it unless a *cis-trans* test has actually been conducted. As an analogy, the utility of atom as a unit did not decrease with the discovery of electron, proton and neutron; similarly, the utility of gene as a unit of heredity should not be undermined with the availability of terms like *cistron, recon* and *muton.* Gene is and will, therefore, continue to be used as an important term in genetics.

At the intragenic level, with the help of recombination studies, micromaps have been prepared for different genes in the same fashion as chromosome maps for different chromosomes were initially prepared at the intergenic level. A gene was thus, further resolved into smaller units called mutational sites. White eye locus and lozenge locus in *Drosophila* have thus each been resolved into a number of mutational sites. Hundreds of other genes in large number of organisms have similarly been mapped.

It should be realized that at a particular stage of research, the number of mutational sites will be minimum, due to experimental limitations of handling limited population. As experimental techniques are refined, better resolution is possible and the existing mutational sites are further resolved into more sites. As mentioned above, the smallest unit capable of undergoing recombination is called *recon. A recon* is further subdivisible into units of mutations called *mutons* and several mutons in a *recon* will not be separable due to recombination. Since mutons in a *recon* can not be identified due to recombination test, they can be recognized through the changes in amino acid sequence of a polypeptide chain synthesized by, or through the nucleotide sequence of, the *cistron* in question. Since a mutation can take place by single base replacement, a single nucleotide pair is the ultimate limit of *muton.*

Therefore, cistron, recon and muton are the units, in the descending order of size and since gene is not synonymous to *cistron,* as explained above, it will be at the top of this hierarchy. Thus a gene can consist of several *cistrons,* a *cistron* of several *recons* and a *recon* of several *mutons.* A *recon* and a *muton* may be of the same size, so that in such a case *recon* may not consist of several *mutons.*

8

Role of Biotechnology

In recent years, techniques for manipulating prokaryotic as well as eukaryotic DNA have witnessed a remarkable development. This has allowed breakage of a DNA molecule at two desired places to isolate a specific DNA segment and then insert it in another DNA molecule at a desired position. The product thus obtained is called recombinant DNA and the technique often called genetic engineering.

Using, this technique we can isolate and clone single copy of a gene or a DNA segment into an indefinite number of copies, all identical. This became possible because, bacteria, phages and plasmids reproduce in their usual manner even after insertion of foreign DNA, so that the inserted DNA will also replicate faithfully with the parent DNA. This technique is called 'gene cloning'. With this technique, genes can be isolated, cloned and characterized, so that the technique has led to significant progress in all areas of cell and molecular biology. More recently, polymerase chain reaction (PCR) involving a thermostable DNA polymerase (Taq polymerase) has been used to obtain millions of copies of a DNA segment of choice. This PCR technique may eventually replace gene cloning in certain areas of research in the field of genetic engineering. In this chapter, we will discuss some details of these techniques involving production of recombinant DNA and the cloning, and amplification of DNA segments.

Restriction Enzymes for Cloning

For cloning of DNA, often we need to cut the DNA at specific sites, which are recognized and cleaved by specific enzymes (endonucleases, which cleave DNA at internal sites), described as restriction enzymes. These restriction enzymes recognize short sequences of double stranded DNA as targets for cleavage. Different enzymes recognize different, but specific sequences, each ranging in length from 4 to 8 base pairs. Each enzyme is named by a three letter (or four letter) abbreviation (letters are italicized) that identifies its origin. Roman numerals (I, II, III, etc.) are added to distinguish several enzymes with same origin. For instance *EcoRI* is derived from *E. coli* and *Hpal* is derived from *Haemophilus parainfluenzae* (Table). By

Recognition Sequences of Several Restriction Endonucleases (Arrow Indicates the Position of Cut).

Microbial origin	*Enzyme*	*Recognition site (two fold symmetry)*
1. *Escherichia coli*	*EcoRI*	5'GAA TTC 3'
		CTT AAG 5'
		↑
		↓
2. *Hemophilus influenzae*	*Hincll*	5' GTPy PuAC 3'
		3' CAPu PyTG 5'
		↑
		↓
	HindIII	5' AAG CTT 3'
		3' TTC GAA 5'

Contd...

Microbial origin	*Enzyme*	*Recognition site (two fold symmetry)*
		↑ ↓
3. *Hemophilus parainfluenzae*	*HpaI*	5' GTT AAC-3' 3' CAA TTG 5' ↑ ↓
	HpaII	5' CC GG 3' 3' GG CC 5' ↑ ↓
4. *Hemophilus aegyptius*	*HaeIII*	5' GG CC 3' 3' CC GG 5' ↑ ↓
5. *Moraxella spp.*	*MspI*	5' CGGG 3' 3' GG GC 5' ↑ ↓
6. *Proteus vulgaris*	*PvuI*	5' CGA TCG 3' 3' GCT AGC 5' ↑ ↓
7. *Bacillus amyloliquefaciens*	*BamHI*	5' GGA TCC 3' 3' CCT AGG 5' ↑ ↓
8. *Streptomyces albus*	*SalI*	5' GTC GAC 3' 3' CAG CTG 5' ↑ ↓

Contd...

Microbial origin	*Enzyme*	*Recognition site (two fold symmetry)*
9. *Providencia stuartii*	*PstI*	5' CTG CAG 3' 3' GAC GTC 5' ↑ ↓
10. *Sterratia marcescens*	*Small*	5' CCC GGG 3' 3' GGG CCC 5'

locating the positions of the cleavage sites of a number of restriction enzymes in a DNA segment, restriction maps can be prepared.

Besides cleavage, modification ir the form of methylation is also brought about by some enzymes called modification enzymes (sometimes also called methylases). This methylation distinguishes genes in different states of functioning. There are also enzymes which perform the function of restriction (cleavage) and modification (methylation). Based on these attributes restriction enzymes have been grouped into two classes :

(i) type II restriction enzyme systems (*e.g. EcoRI),* which have separate enzymes for modification and restriction and

(ii) type I *(EcoK* and *EcoB)* and type III *(EcoPl, Eco 15)* enzyme systems, in which same enzyme possesses both activities (bifunctional), although the restriction and modification sites differ in position.

Of the above two classes of restriction enzymes, type II enzymes are most important for cloning purposes. Some enzymes introduce staggered cuts, others generate blunt ends. Enzymes with 4bp target sites are used when frequent cuts are desired to get small DNA fragments and those with 8bp are used when rare cuts are desired to get long DNA segments. Otherwise majority of enzymes used have 6bp target sites. Some of them can cleave both methylated as well as unmethylated targets, but majority of them cleave only unmethylated targets. Details of some of these enzymes are listed in Table.

Techniques used in Recombinant DNA Technology

Separation of DNA Fragments Using Agarose or Polyacrylamide Gel Electrophoresis : When genomic DNA, extracted from any tissue of a plant or animal species, is digested with a restriction enzyme, it is cleaved into segments. The, segments of different sizes can be separated through gel electrophoresis before a molecular probe is used to detect the segments which have sequences similar to those in the probe (for molecular probes, see later in this chapter). Gel electrophoresis involves movement of fragments or molecules from a well created on one edge of the gel. The gel may be a cylinder or a slab (usually a slab for cloning expt.), about 10 cm long and 0-5 cm thick. The rate of movement of fragments is inversely correlated with the size of fragments or molecules, so that heavier fragments will remain closer to the site of loading and the lighter fragments will move away (Fig.). Fragments of different sizes will appear as bands on the gel and can be examined or isolated for further study. More often agarose gels are used, but for separation of fragments differing by few base pairs, polyacrylamide gels are used. Polyacrylamide gels are more commonly used for DNA sequencing experiments.

Separation of Large DNA Molecules (Whole Chromosomes) using PFGE : The technique of gel electrophoresis is used for separation of DNA molecules of different sizes. However, DNA molecules of large size could not be handled in this technique. In recent years, a new technique called pulsed field gel electrophoresis (PFGE) has been used, for separation of large sized DNA molecules, sometimes representing whole chromosomes. Using this technique, separation of DNA molecules belonging to each of the 16 individual chromosomes of yeast has become possible (Fig.). In this technique, short pulses of electricity are used in two different directions and DNA is embedded and used in the form of agarose plugs to avoid fragmentation of large DNA molecules. Using the technique of PFGE and a more refined technique CHEFE (contour clamped homogeneous electric field electrophoresis), genomes of several fungi could be resolved into chromosomal bands and used for mapping of DNA sequences on specific chromosomes.

Southern, Northern and Western Blotting : A mixture of DNA, RNA or protein fragments can be separated by gel electrophoresis and the separated bands can be stained and visualized directly in the gel. However, to confirm the identity of these bands or to find similarity of one or more of these bands with a known and available molecular probe, it is possibie to hybridize these bands with a labelled probe. Although hybridization of separated DNA fragments with a probe of small size (10-20 base) can be conducted within the gel (in-gel hybridization), to facilitate this hybridization with relatively bigger probes, the bands are often transferred to a nitrocellulose membrane through a technique described as blotting. When DNA bands are thus blotted, it is called Southern blotting

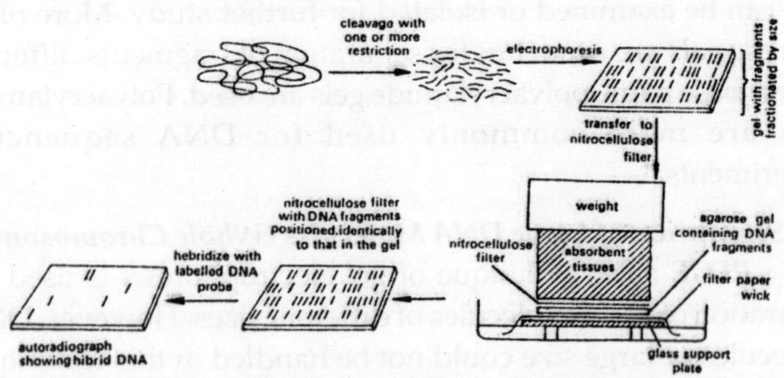

Southern blotting method of analysing DNA segments that share homology with a nucleic acid probe.

(after the name of E.M. Southern); when RNA bands are thus transferred it is described by the jargon term Northern blotting and similarly when protein bands are transferred, the technique is described as Western blotting.

Southern Blotting: For Southern blotting, DNA sample is first digested with a restriction enzyme and digested sample is gel electrophoresed. The DNA bands in the gel are denatured into single strands with the help of an alkali solution. Subsequently, the gel is laid on top of a buffer saturated filter paper, placed on a solid support (*e.g.* glass plate), with its two edges immersed in the buffer. A sheet of nitrocellulose membrane is placed on top of the gel and a stack of many papers (paper towels) on top of this membrane. A weight of about 0.5 kg. is placed on top of paper towels.

The buffer solution is drawn up by filter paper wick, and passes through the gel to the nitrocellulose membrane and finally to the paper towels. While passing through the gel, the buffer carries with it single stranded DNA, which binds on to the nitrocellulose membrane, when the buffer passes through it to the paper towels. After leaving this arrangement for a few hours or overnight, paper towels are removed and discarded. The nitrocellulose membrane with single stranded DNA bands blotted on to it, is baked at 80°C for 2-3 hours to fix the DNA permanently on the membrane. This membrane now has a replica of DNA bands from agarose gel, and can be used for hybridization with labelled DNA or RNA probe. The membrane may then be washed to remove any unbound DNA, and X-ray film is exposed to the hybridized membrane to get autoradiographs, The above steps involved in Southern blotting are shown in Figure.

Northern Blotting: Initially the technique of Southern blotting used for DNA transfer from gel to the membrane could not be used for blot-transfer of RNA. Instead mRNA bands from the gel were blot-transferred into a chemically reactive paper, prepared by diazotization of amino-benzyloxymethyl paper.

The technique being related to Southern blotting was called Northern blotting (not after name of any scientist as in Southern blotting). Later, it was shown that mRNA bands can be blotted directly onto nitrocellulose membrane, a technique which has been widely adopted. The mRNA bands blotted onto nitrocellulose membrane can be hybridized with a labelled DNA or RNA probe. The single stranded regions of the probe are removed by nuclease (mungbean nuclease or S-l nuclease), so that quantitative estimation of hybridized mRNA can also be made.

Western Blotting: This technique is used to detect proteins of a particular specificity. Particularly when a transferred gene expresses in transformed cells, the translated product in the form of protein can be identified by this technique. The extracted proteins are subjected to polyacrylamide gel electrophoresis (PAGE) and are then transferred onto nitrocellulose, to which they bind. Nitrocellulose membrane is then used for probing with a specific labelled antibody (antibody will not hybridize with protein, but bind with it). The antibody may be labelled with ^{125}I and the signal is detected again with autoradiography. If radioactive label is not used, bound antibody may be detected by a second antibody tagged with an enzyme.

Dot Blots and Slot Blots : Another variety of Southern of Northern blots is dot blots or slot blots. In dot blots, cloned or pure extracted DNAs to be tested are spotted adjacent to each other on a nitrocellulose membrane. DNA blots thus produced are immobilized and denatured so that they are bound on the membrane as single stranded DNA blots. These blots are then hybridized with radioactively labelled probe (DNA or RNA). The dot, representing a sequence related to probe, will light up by autoradiography (Fig.). The intensity of the dot will indicate the relative concentration of the

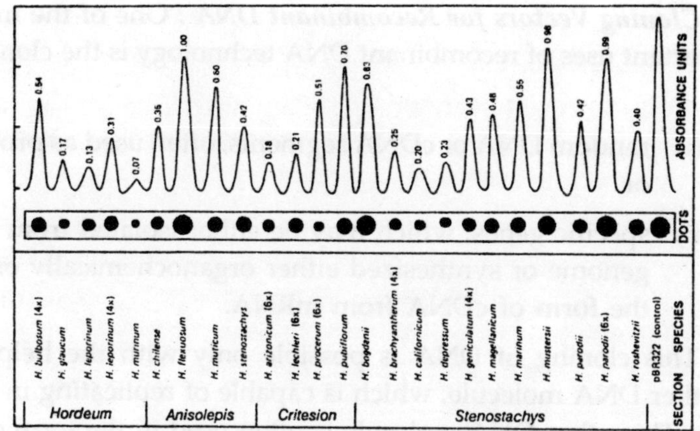

Dot blots from 25 Hordeum species, showing difference in abundance of sequence related to the probe pSC119.

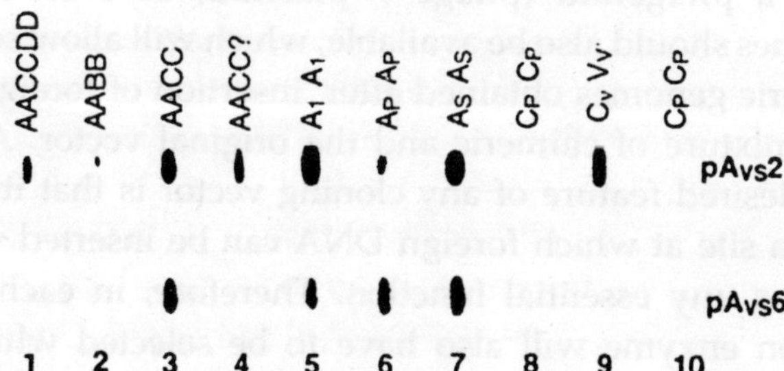

Slot blots from genomic DNA of 10 Avena species showing differences in abundance of sequences related to two probes (pAvs2, pAvs6).

sequence related to the probe in the DNA sample used for a particular dot. No steps involving digestion with enzyme, gel electrophoresis or transfer from gel to membrane are needed in this technique. Therefore, it is considered much more convenient, when no restriction sites are needed to be studied.

Sometimes, an apparatus in used for placing spots on the membrane, through slots made in this equipment. The spots made thus are in the form of oblong slots rather than round blots. These slots are used just like dot blots and are described as slot blots (Fig.).

Cloning Vectors for Recombinant DNA : One of the most important uses of recombinant DNA technology is the cloning of :

(i) random DNA or cDNA segments, often used as probes or

(ii) specific genes, which may be either isolated from the genome or synthesized either organochemically or in the form of cDNA from mRNA.

This cloning of DNA is possible only with the help of another DNA molecule, which is capable of replicating in the host. This other DNA molecule is often used in the form of a vector, which could be a plasmid, a bacteriophage, a derived cosmid, a phagemid (phage + plasmid) or even a virus. Techniques should also be available, which will allow selection of chimeric genomes obtained after' insertion of foreign DNA from a mixture of chimeric and the original vector. Another critical desired feature of any cloning vector is that it should possess a site at which foreign DNA can be inserted without disrupting any essential function. Therefore, in each case a restriction enzyme will also have to be selected which will cause a single break. Sometimes vectors are modified by inserting a DNA segment to create unique site(s) for one or more restriction enzymes to facilitate its use in gene cloning. This inserted DNA is sometimes called 'polylinker'.

Plasmids as Vectors

Plasmids are defined as autonomous elements, whose genomes exist in the cell as extrachromosomal units. These plasmids can be single copy plasmids that are maintained as one plasmid DNA per host chromosome or multicopy plasmids, which are under relaxed replication control, thus permitting their accumulation in very large numbers (up to 1000 copies per cell). These latter type of plasmids are used as cloning vectors, due to their increased yield potential.

Circular plasmid DNA which is used as a vector, can be cleaved at one site with the help of an enzyme to give a linear DNA molecule. A foreign DNA segment can now be inserted, by joining the ends of broken circular DNA to the two ends of foreign DNA, thus regenerating a bigger circular DNA molecule (Fig.), that can now be separated by gel electrophoresis on the basis of its size.

Selection of chimeric DNA is also facilitated by the resistance genes which the plasmid may carry against a certain antibiotic. If a plasmid has two such genes conferring resistance against two antibiotics and if the foreign DNA insertion site lies within one of these two genes, then the chimeric vector loses resistance against one antibiotic, the gene for which has foreign DNA inserted within its structure.

In such a situation, the parent vector in bacterial cells can be selected by resistance against two antibiotics and the chimeric DNA can be selected by retention of resistance against only one of the two antibiotics.

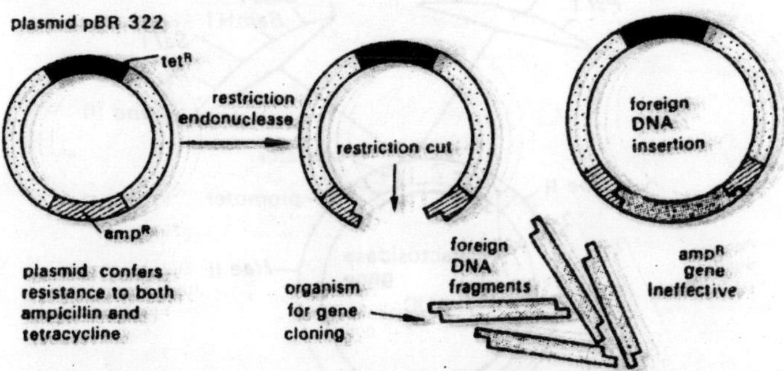

The concept of cloning a foreign DNA segment in plasmid vector.

The naturally occurring plasmids may not possess all the above and other essential properties of a suitable cloning vector. Therefore one may have to restructure them by inserting genes of relaxed replication and/or genes for antibiotic resistance. This has actually been done and suitable plasmid vectors have been obtained. Following are some popular examples.

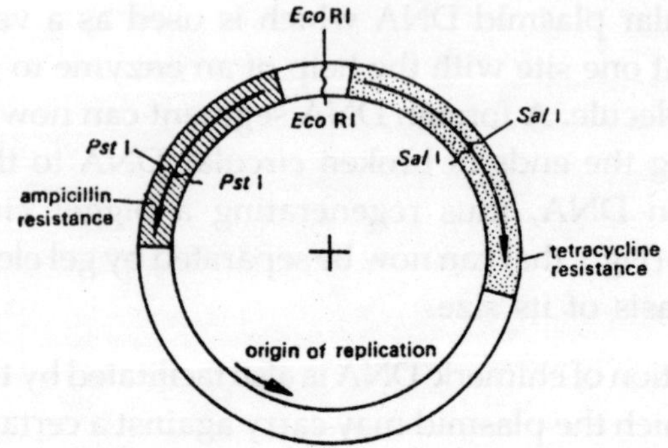

Plasmid pBR322 and its major characteristic features.

pBR322 and pUC Vectors : One of the standard cloning vectors widely used in gene cloning experiments is pBR322 (derived from *E. coli* plasmid ColEl), which is 4,362 bp DNA and was derived by several alterations in earlier cloning vectors (Fig.). A characteristic feature of pBR322 is, that it contains

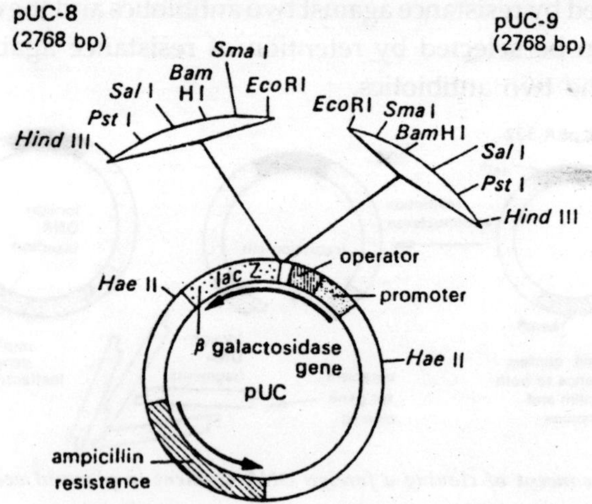

Plasmid pUC and its major characteristic features.

resistance against two antibiotics due to two genes, one for resistance against ampicillin and the other for resistance against tetracyclin. Another series of plasmids that are used as cloning vectors belong to pUC series (named after the place of their initial preparation *i.e.* University of California). These cloning vectors are available in pairs with reversed orders of restriction sites relative to *lacZ* promoter; pUC8 and pUC9 make one such pair (Fig.). The pUC vectors carry a *Hindi* fragment of *lacZ* gene. The product of *Hindi* fragment *i.e.* α peptide complements with β peptide produced by the defective gene in the host to produce functional β galactosidase; the phenomenon is described as α complementation. For multiplication of recombinant pUC vector, a host is used, which has a defective *lacZ* gene producing only β peptide of β-galactosidase. The presence of β galactosidase, due to α complementation, is detected by white colonies as against blue colonies produced by non-recombinant vector in presence of an inducer, IPTG and the dye Xgal.

As discussed above, in pBR 322, foreign DNA is inserted at a site located in one of the two genes for resistance against antibiotics, so that it will inactivate one of the two resistance genes. The insert bearing plasmids can be selected by their ability to grow in a medium containing only one of the two antibiotics and by their failure to grow in a medium containing both the antibiotics. The plasmids carrying no insert on the other hand, will be able to grow in media containing one or both the antibiotics. In this manner, the presence of *Hindi* fragment of *lacZ* gene in pUC and resistance genes against ampicillin and tetracyclin in pBR322 allow selection of *E. coli* colonies transformed with plasmids carrying the foreign DNA segment.

Ti and Ri Plasmids as Vectors for Higher Plants : Among higher plants, Ti plasmid of *Agrobacterium tumefaciens* and Ri plasmid of *A. rhizogenes* are the best known vectors. T-DNA, from Ti or Ri plasmid of *Agrobacterium*, is considered to be a very potential vector for cloning experiments with higher plants. This will involve the following steps : (i) foreign DNA has to be first cloned into T-DNA of Ti or Ri plasmid, which is then transferred to *Agrobacterium*, (ii) modified hybrid T-DNA can be transferred to the genome of plant cells by *Agrobacterium* infection. Recombinant Ti plasmids can also be obtained by initially cloning of a large fragment of Ti plasmid in pBR322 and then transferring it to *Agrobacterium* through conjugation and recombination. These recombinant Ti plasmids can then be used for transformation of higher plants.

Bacteriophages as Vectors

Bacteriophages provide another source of cloning vectors. Since usually, a phage has a linear DNA molecule, a single break will generate two fragments, which are later joined together with foreign DNA to generate a chimeric phage particle (Fig.). The chimeric phage can be isolated after a lytic cycle. The use of phage particles as vector imposes a limitation on the size of foreign DNA, which can be cloned, because the capacity of phage head is only limited, and if the size of foreign DNA is too long, size of phage DNA may not be accommodated in phage head. In order to overcome this problem, those segments of DNA, which did not contain essential genes could be removed. Such a technique has been followed in phage lambda (λ) to create a smaller vector genome having single restriction site for the enzyme *EcoRI.* Since the reduced size also fails to be adequately packed in phage head (there is also a requirement of a minimum size of DNA), this automatically provides a selection method, in which only the chimeric particles will be obtained in the phage progeny, and vector particles lacking cloned segment will be eliminated due to its reduced size.

Plasmid vectors are often used for cloning DNA segments of small size (up to 10 kilobases). However, while preparing a genomic library in a eukaryote, the cloned fragments should be large enough to contain a whole gene. This will also allow cloning of the whole genome into a number (of clones), which will not be unreasonably large and therefore can be screened without serious difficulty. These properties and other requirements of cloning whole genome in eukaryotes are fulfilled by the phage lambda, cosmid or phagemid vectors, the former permitting cloning of segments up to 20-25kb long (kb = kilobases) and latter accommodating segments up to 45kb long. Phage lambda *(λ),* however, is easier and more efficient for making genomic and cDNA libraries. Some of the vectors derived from lambda and M13 phages are described in the following text.

λ gt10 and λ gt11: λ gt10 is 43kb double stranded DNA for cloning fragments that are only up to 7kb in length. The insertion of DNA inactivates *cl*⁺ (produces repressor, necessary for lysogeny) gene generating a *cl*⁻ bacteriophage. Nonrecombinant λ gt10 is *ct*⁺ and forms a cloudy plaque on appropriate *E. coli*

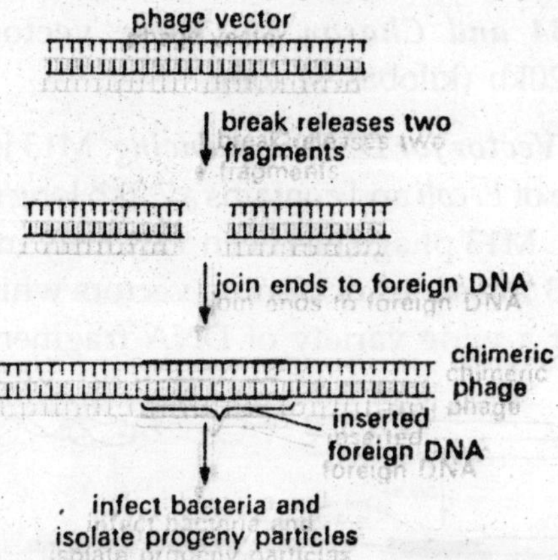

Cloning of foreign DNA in a nonessential region of phage vector.

host (due to lysogeny, no lysis), while recombinant c⁻ λ gt10 forms clear plaques (due to lysis) permitting screening of recombinant plaques. Recombinant lgt10 plaques can thus be easily selected.

λ gtl1 is a 43.7kb double stranded λ phage for cloning DNA segments, which are less than 6kb in length (usually for cDNA). It differs from λ gtl0 in being an expression vector so that foreign DNA can be expressed as β galactosidase fusion proteins. The recombinant λ gtll becomes *gal⁻*, while nonrecombinant λ gt11 remains *gal⁺*, so that on an appropriate *E. coli* host, in the presence of IPTG (inducer) and Xgal (substrate), recombinant phage *(gal⁻)* will form white colonies or clear plaques and non-recombinant phage *(gal⁺)* will form blue colonies permitting screening (due to α - complementation).

EMBL3 and EMBL4: In these vectors a central non-essential part of the phage, 44kb long, can be replaced by a foreign DNA which may be as long as 20-23 kb. Therefore, they are called replacement vectors. Non-recombinant DNA is too small for packaging, so that there is automatic selection against non-recombinant phages (Fig.).

Charon 34 and Charon 35: These vectors will accept fragments 9-20kb (kilobases) long.

M13 as a Vector for DNA Sequencing: M13 is a filamentous bacteriophage of *E. coli* and contains a 7.2kb long single stranded circular DNA. M13 phage has been variously modified to give rise to a MP13 mp series of cloning vectors which can be used for cloning of a wide variety of DNA fragments particularly

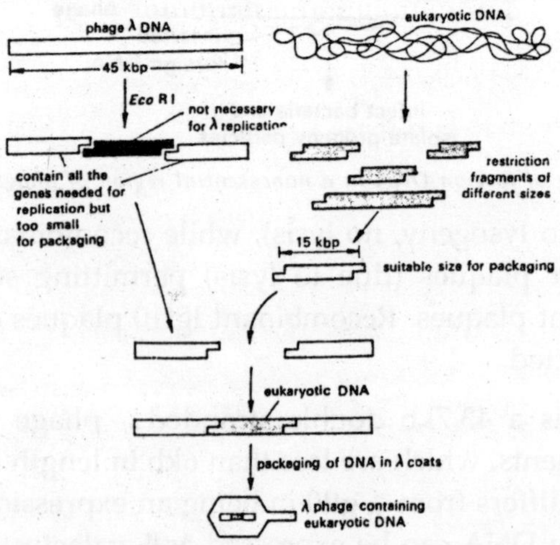

Use of lambda (λ) phage as vector for cloning eukaryotic DNA fragments.

for the purpose of. sequencing through Sanger's method of dideoxy chain termination.

Cosmids and Phagemids as Vectors

Cosmids are plasmid particles, into which certain specific DNA sequences, namely those for *cos* sites are inserted. Since these *cos* sites enable the DNA to get packed in lambda particle, cosmids allow the packaging of DNA in phage particle *in vitro,* thus permitting their purification. Like plasmids, these cosmids perpetuate in bacteria and do not carry the genes for lytic development. Phagemids are prepared artificially by combining features of phages (*e.g.* M13) with plasmids, as the name suggests. One such phagemid, which is commonly used in molecular biology laboratories is pBluescript II KS, which is derived from pUC19, and is 2961 base pairs long. This is an expression vector, since the cloning site is flanked by T_3 and T_7 promoters to be read in opposite directions.

Plant and Animal Viruses as Vectors

A number of plant and animal viruses have also been used as vectors both for introducing foreign genes into cells and for gene amplification and expression in host cells. Following are some of the viruses that are commonly used as vectors : (i) plant viruses (cauliflower mosaic virus or CaMV, TMV, and geminiviruses); (ii) animal viruses (baculovirus, adenoviruses, papovaviruses, herpesvirus, retroviruses).

Transposons as Vectors

Transposons of higher plants (*e.g. Ds, Ac* or *Mul* of maize) and transposons of *Drosophila* (P elements) are also used as vectors.

Artificial Chromosome Vectors for Cloning Large DNA Segments : We know that in plasmid vectors, sequences up to 10-15kb, in lambda (A.) phage vectors, sequences up to 22kb and in cosmid vectors, sequences up to 40kb can be cloned. For

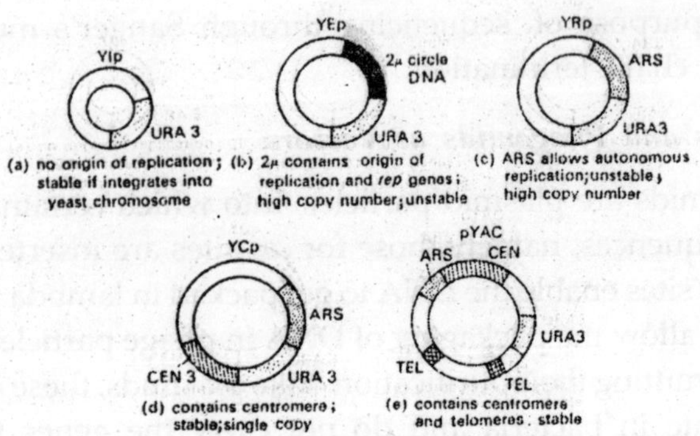

Five different yeast plasmid vectors.

cloning sequences much bigger in size (> 100kb), artificial chromosome vectors have been designed, by combining the DNA sequences necessary for replication, segregation and maintenence of these chromosomes. These artificial chromosomes have now been constructed for yeast (yeast artificial chromosomes or YACs), mammals including humans (mammalian/human artificial chromosomes or MACs/HACs) and bacteria (bacterial artificial chromosomes or BACs).

Yeast Artificial Chromosomes (YACs): Yeast artificial chromosome vectors have been constructed utilizing the yeast centromere sequences (CEN), telomere sequences (TEL), autonomous replicating sequences (ARS) and other necessary DNA sequences. They allow cloning of sequences that are several hundred kilobase pairs long. The sequence to be cloned can be ligated to YAC vector, which can then be multiplied as a linear artificial chromosome in yeast cells. These long DNA molecules for cloning can be generated by cleavage with certain restriction enzymes or by partial cleavage, and allow construction of comprehensive libraries (with large DNA segments) in microbial hosts. These vectors along with other yeast plasmid vectors are shown in Figure.

Mammalian Artificial Chromosomes (MACs): In 1997 and 1998, utilizing the human centromere (tandem arrays of alpha DNA) and telomere sequences along with high molecular weight (HMW) genomic DNA, human artificial chromosomes (HACs) were constructed for the first time. These MACs/ HACs were introduced in cultured cells and their successful multiplication, inheritance and maintenance was demonstrated (loss rate=<0.5%). This successful creation and use of human artificial chromosomes is a significant development towards the MAC-based vector system for gene therapy, since they will allow long term maintenance of therapeutic DNA in living cells.

Bacterial Artificial Chromosomes (BACs): YAC and MAC vectors, though, are very useful for a variety of purposes (particularly for cloning fragments bigger than 1000kb), but are too big and are difficult to use because they need to be multiplied in yeast and mammalian cells respectively. Therefore, a user friendly bacterial cloning system based on *E. coli* F factor has been designed, which can be used for cloning fragments as long as 300-350kb in bacterial cells. The construction of BACs involved the use of all essential regulatory DNA sequences of F factor, a chloramphenicol resistance gene and a segment having the cloning site, etc. These vectors are being extensively utilised in plant and animal systems for a variety of purposes.

Construction of Chimeric DNA

Chimeric DNA molecules are produced by inserting a foreign DNA segment into a DNA molecule of a vector. These

(*Eco*RI recongnizes 5′GAA3′)

5′ AACCACAGAATTCACACC 3′
3′ TTGGTGTCTTAAGTGTGG 5′

5′ AA CCACAG AATT CACACC 3′
3′ TT GGTGTCTTAA GTGTGG 5′

Staggered cleabager of a double stranded DNA fragment, producing single stranded sticky ends.

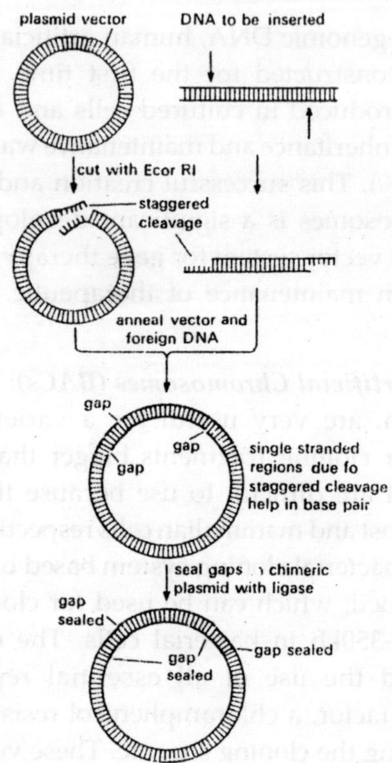

Cloning of a DNA segment in a plasmid vector (e.g. pBR322) by staggered cleavages, both in vector DNA and foreign DNA to be cloned.

chimeric DNA molecules are the means of cloning genes or other small DNA sequences. The construction of chimeric DNA, as we know now, involves joining of the broken ends of the vector with the two ends of the sequence to be cloned. There are atleast four methods available for the construction of this chimeric DNA.

Cloning Fragments with Homologous Cohesive Ends (Palindromes and Staggered Cleavage) : This is the most common (but least efficient) method, where due to the presence of a palindromic sequence, a restriction enzyme causes staggered cuts producing short complementary single stranded sticky ends (a palindromic sequence is one, which has complementary sequences at the two ends of a single strand, *e.g.* ATC—GAT). This can be illustrated using the example of enzyme *EcoRl,* which cuts the sequences at specific positions (Fig.). When another DNA molecule is similarly cut by the same enzyme, similar sticky ends having same sequences in the single stranded ends will be produced, so that when it is mixed with the previous molecule similarly treated, the two will anneal producing a chimeric DNA or chimeric plasmid, if plasmid DNA is involved (Fig.). Enzyme DNA ligase helps in joining the cut ends of two molecules.

This technique has the advantage of regenerating two restriction sites (*e.g., EcoRl* sites) in the chimeric DNA, so that the foreign DNA segment can be retrieved rather easily from the cloned copies of chimeric DNA by cleavage again with the same enzyme. There are also following disadvantages with this technique:

(i) The two cleaved ends of a vector or of a foreign DNA may join end to end (self annealing) before the foreign DNA gets inserted. Multiple copies of the foreign DNA may also be inserted in tandem order. Therefore, while isolating chimeric DNA, one will have to select chimeric molecules each having only a single insert, by separating molecules of different sizes by gel electrophoresis. Problem of self annealing may be partly overcome by treating the cleaved vector DNA (and not the foreign DNA) with an alkaline phosphatase so that the ends will not self ligate leading to recircularization of vector.

(ii) The foreign DNA may be cloned in either one of the two orientations, which can be detected only by restriction mapping.

(iii) The recognition site, particularly in the sequence to be cloned, may not lie at a convenient position, so that sometimes only a part of the desired segment will be inserted. For the above reasons, cloning through the use of staggered cohesive ends is avoided, whenever possible; instead directional cloning discussed below is employed.

Directional Cloning using Fragments with Heterologous Ends : In this method, the vector as well as the foreign DNA is cleaved by two enzymes thus removing a segment of the DNA lying between the cleavage sites of two enzymes. This eliminates self-ligation and if staggered ends are produced, the efficiency of cloning is high. Further, in this method one copy of the insert DNA is oriented in a defined direction with respect to the vector, hence the name 'directional cloning'. Problems in this method mainly arise, when after enzyme treatment, some of the DNA fragments are produced due to cleavage by one enzyme instead by both the enzymes. This means that the cleaved vector will consist of a mixture of fragments, so that the fragments produced due to cleavage by both enzymes will have to be purified.

Adding Poly dA at the 3' Ends of the Vector and Poly dT at the 3' Ends of DNA Clone : In this method DNA can be cut at the desired position both in the vector and in the clone, without staggered cleavage. This produces blunt ends. Using precursor dATP, poly dA is added at both the 3' cut ends of the vector with the help of the enzyme terminal transferase. Similarly, using precursor dTTP, poly dT is added at both the 3' cut ends of (he DNA sequence to be cloned.

The vector and clone can then the joined by annealing the poly dA with poly dT tails and then ligating them using DNA ligase enzyme (Fig.). In this technique, there is no chance of reannealing between the two cut ends of the same DNA molecule, so that one of the disadvantages of the first method has been overcome in this case. However, in this technique it will not be easy to retrieve the cloned DNA, because the recognition site of the enzyme has been lost due to insertion of poly dA and poly dT. But if poly dG and poly dC are used instead (which serve the same purpose), cloned sequence can be retrieved, since poly dG : poly dC regenerate the recognition site for the enzyme *Pstl* or Small (Table).

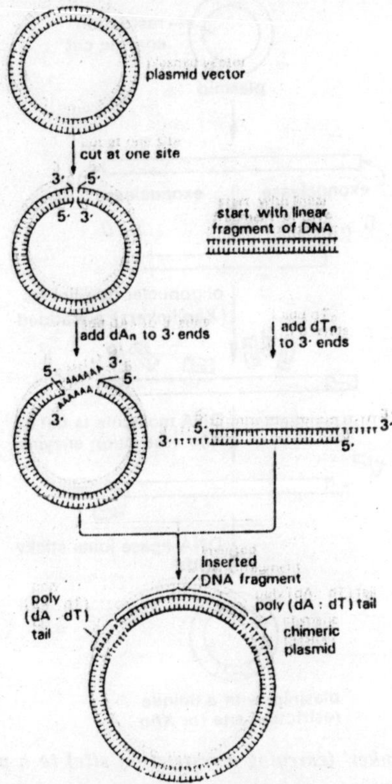

Cloning of DNA segment in a plasmid vector by poly dA: poly dT tailing technique.

Blunt end Ligation by T4 DNA Ligase

In this technique, a restriction enzyme is used to cut a duplex DNA at the same place in both the DNA strands. The broken ends are then used for joining with the two blunt ends of another DNA molecule irrespective of the sequences present at the broken ends of the two DNA molecules. The T4 DNA ligase is used for this joining reaction. The disadvantage of this technique is that any two broken ends may join including those belonging to the same DNA molecule. This leads to the production of a variety of products and one will have to select the desired product from a mixture of products.

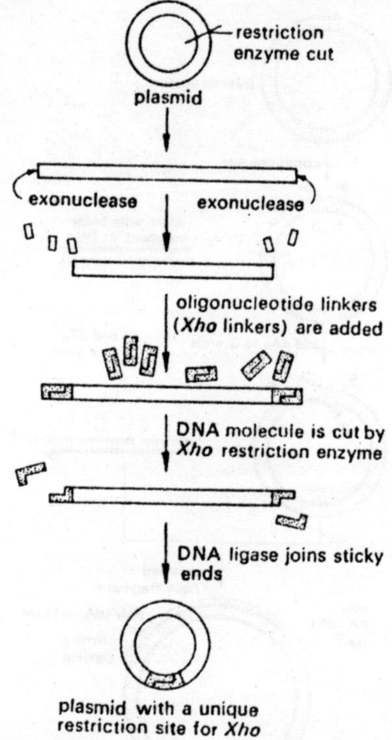

Addition of a 'linker' (carrying a restriction site) to a vector molecule.

The blunt end ligation is also used for developing a method in which the cloned DNA can be easily retrieved whenever required. This method makes use of short DNA duplexes ('linkers'), that contain *EcoRI* palindrome or some equivalent palindrome, which being small in size, can be synthesized chemically. These 'linkers' can be linked to the blunt ends of vector DNA or of an insert by blunt end ligation (Fig.). This will allow the creation of an *EcoRI* site in the linker region of the vector. Therefore, with this method it is now possible to insert a foreign DNA segment at a particular site in the linker region of the vector and then retrieve this foreign DNA segment whenever necessary.

Cloning in Bacteria and Eukaryotes

Cloning in Bacteria

Bacteria can be classified into (i) gram negative bacteria *e.g. Escherichia coli, Pseudomonas, Rhizobium* and (ii) gram positive bacteria *e.g. Bacillus subtilis* and *Streptomycetes* sp. Ordinarily plasmids can not maintain themselves in stable form in both classes of bacteria and, therefore, plasmid vectors for two classes usually differ. While for gram negative bacteria, plasmids, phages and cosmids are known as vectors, for gram positive bacteria, only plasmids are known to do this job.

Cloning in Eukaryotes

Since chromosomes found in the nucleus in eukaryotes are separated from the rest of cell through nuclear membrane, and since many of the genes are split genes with exons and introns, genetic engineering with eukaryotes requires new methods and tools. When eukaryotic genes are cloned in prokaryotes, the split genes can not be correctly expressed, because prokaryotes do not have the equipment necessary for splicing out the RNA transcribed from the introns of a gene.

Among eukaryotes, DNA cloning has been done in yeast, mouse and to some extent even in some higher plant species. In yeast, a plasmid called 2 μ DNA (63bp) is found, which is an appropriate cloning vehicle. An efficient transformation method is also available, which involves protoplast production and PEG (polyethylene glycol) directed introduction of DNA. In animal cells like mouse cells, special animal viruses were used as cloning vehicles. Simian virus 40 (SV40) is one such virus, in which globin gene could be integrated. This gene integrated in SV40 could be transcribed and translated in mouse kidney cells.

Molecular Probes

Molecular probes are small DNA segments (genomic DNA, cDNA or synthetic oligonucleotides) or RNA segments (often synthesized on DNA template). They recognize complementary sequences in DNA or RNA molecules and thus allow identification and isolation of these specific DNA sequences from an organism. Antibodies are also occasionally used as probes to recognize specific protein sequences. These probes are also frequently used for a variety of other purposes including diagnosis of infectious diseases, identification of food contaminants, variety of microbiological tests, forensic tests (*e.g.* fingerprinting of murderers or rapists), etc. Probes can

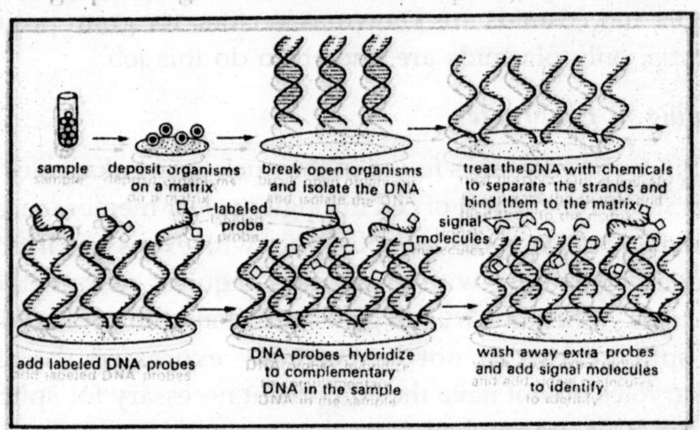

Use of labelled DNA probe for identification of homologous DNA sequence through DNA hybridization.

also be used to identify strains of an organism *e.g.* varieties of a crop species {a plant breeder likes to have quick test to identify his variety to maintain his patent or plant breeder's right (PBR), so that a competitor may not use it in another name leading to infringement of rights}.

DNA/RNA probe assays are faster and sensitive, so that many conventional diagnostic tests for viruses and bacteria involving culturing of the organism, are being fast replaced by antibody and DNA probe assays. While culture tests can take days or even months, molecular probe assays can be performed within few hours or minutes. Therefore, the use of DNA probes has become today's most sophisticated and sensitive technology for a variety of uses involving biological systems both in basic and applied studies including their commercial use.

Labelling of Probes

The detection of homologous sequences after hybridization with a probe is like finding a needle in the haystack. Therefore, for the success of DNA probe assay, it is necessary to develop simple, safe and sensitive techniques for their use. As probes transmit no signal of their own, they have to be (i) labelled either with radioactive isotopes (*e.g.* ^{32}P dCTP) or (ii) with non-radioactive signal molecules (*e.g.* biotin or digoxigenin) without impairing the hybridization ability of these probes. These signal molecules may include fluorescent antibodies, enzymes that produce colour changes in dyes and chemiluminescent catalysts (Fig.).

Radioactive Labelling: After hybridization with radioactively labelled probe, hybrids are detected by autoradiography (Fig.). 32P has the advantage over other radioisotopes, since it has high specific activity. However, in general, radioisotopes have some disadvantages. They are difficult to handle and expensive to dispose off. Detection by autoradiography, while sensitive, may take a long time if there are few counts in the hybrid. Furthermore, radioisotopes have a short half life (*e.g.* 32P has a half-life of 14.3 days) and therefore experiments should be completed preferably within one half-life.

Non-radioactive Labelling: Recent advances in nucleic acid technology now offer alternatives to radioactively labelled probes. One procedure that is becoming increasingly popular is biotin labelling of nucleic acids. After hybridization and washing, detection of hybrids is done by a series of cytochemical reactions which finally give a blue colour whose intensity is proportional to the amount of biotin in the hybrid. There are several advantages of using biotinylated probes. For example, these assays employ non-toxic materials, whose half-life is longer. These probes can be prepared in advance in bulk and stored at - 20°C for repeated uses.

Digoxigenin is another chemical derived from plants and used for non-radioactive labelling of probes. An antibody associated with an enzyme (anti- digoxigenin-alkaline phosphatase conjugate) is used for the detection of the presence of digoxigenin. The probe may be labelled with digoxigenin-11-dUTP supplied with a digoxigenin kit (these kits are available from any commercial firm, *e.g.* Boehringer Mannheim). The labelled and denatured probe may be used for hybridization with denatured DNA on Southern blots or for *in-situ* hybridization. After hybridization the membrane or the slide may be washed and the membrane or the slide is transferred into detection buffer containing 20 m*g/ml* of anti-digoxigenin fluorescein and 5% (w/v) BSA (bovine serum albumin). The system is incubated for lh at 37°C, and then the membrane/ slide is washed in detection buffer three times (8 min each at 37°C). Alkaline phosphatase activity can be detected using 0.17 mg/ml BCIP (5-bromo 4-chloro 3-indolyl phosphate) and 0.33 mg/ml NBT (nitroblue tetrazolium) as dye substrate.

Applications of Molecular Probes

Molecular probes can be used for both basic and applied studies in the field of molecular biology and biotechnology. These uses include the following: (i) 'Restriction Fragment Length Polymorphisms (RFLPs)' and their uses for preparation of molecular maps and other purposes, (ii) Screening of genomic and cDNA libraries for isolation of genes etc. (see later in the chapter), (iii) Location of genes or DNA sequences on chromosomes, through in *situ* hybridization or ISH, (iv) Use in diagnosis of diseases and DNA fingerprinting.

Construction and Screening of Genomic and cDNA Libraries : In order to isolate one or more related genes from a genome, we like to prepare a mixture of clones each carrying DNA derived either from the genomic DNA or from cDNA (derived from the mRNA isolated from a specific metabolically active tissue of an organism). This mixture may contain thousands of clones, which when derived directly from the genomic DNA are collectively called a genomic library. Similarly when these clones are cDNA, they are collectively called a cDNA library. Construction and use of these libraries will be discussed in this section.

Genomic Library by Shotgun Experiment

Cloning an entire genome in the form of a library of random genomic clones (without identifying them) is often called a shotgun experiment. In this experiment, genomic DNA is extracted, broken into fragments of reasonable size by a restriction endonuclease and then inserted into a cloning vector to generate a population of chimeric vectors (Fig.). A set of fragments cloned in this manner is called a genomic library. Once such a library is available, then clones can be perpetuated indefinitely in a plasmid vector and retrieved whenever needed for a variety of purposes, mainly for identification and isolation of a gene, whenever a specific probe is available.

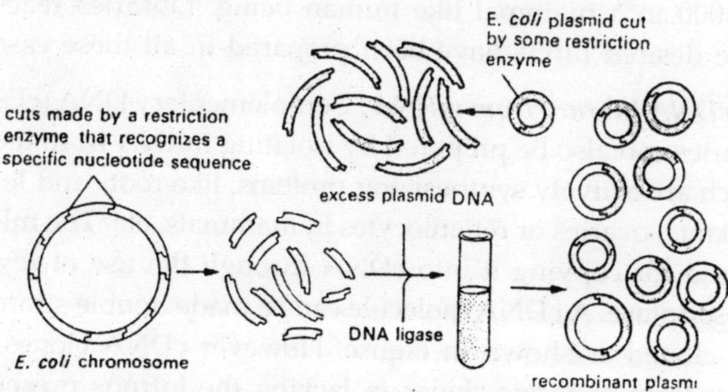

Formation of a genomic library using recombinant DNA technique.

Genomic libraries can be prepared by using a number of restriction endonucleases, one at a time, so that fragments of varying sizes having cuts at different places of the genome will be available. However this may lead to cuts at inconvenient places, including sites within a gene, so that fragments having complete genes will be difficult to obtain. In order to overcome this difficulty, we use the following strategy in the shotgun experiment : (i) We use restriction endonucleases:, which have short (4bp) recognition sequences, so that such a sequence may be frequently distributed. (ii) Conditions are used which give only partial digests, so that a particular restriction site is only occasionally cleaved, and long fragments without having any breaks on recognition sites available within a gene can be easily obtained. This technique of shotgun experiment leads to the construction of a random genomic library, in which all cloned fragments have same sequences at the two ends thus helping in retrieval of a fragment from the vector with the help of the same enzyme.

The number of fragments representing every sequence of the genome increases with genome size. For instance, for a probability level of 99% that all sequences are present in our library of a species, we may need 1,500 cloned fragments for *E. coli;* 4,600 in yeast; 48,000 in *Drosophila melanogaster* and 8,00,000 in a mammal like human being. Libraries reaching these desired limits have been prepared in all these cases.

cDNA Library from mRNA : Complementary DNA (cDNA) libraries can also be prepared by isolating mRNA from tissues which are actively synthesizing proteins, like roots and leaves in plants, ovaries or reticulocytes in mammals, etc. The mRNA is used for copying it into cDNA through the use of reverse transcriptase. A cDNA molecule can be made double stranded, and cloned as shown in Figure. However cDNA clones will differ from genomic clones in lacking the introns present in split genes, but may have the advantage of being expressed in bacteria, which do not have the machinery to process the hnRNA obtained from split genes, into mRNA.

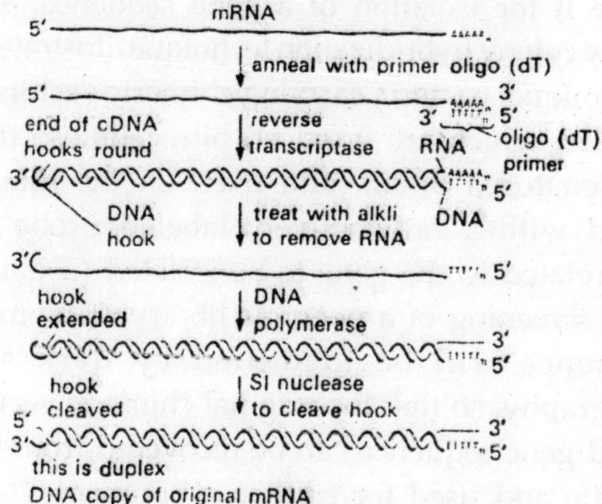

Synthesis of cDNA from mRNA, using reverse transcriptase enzyme.

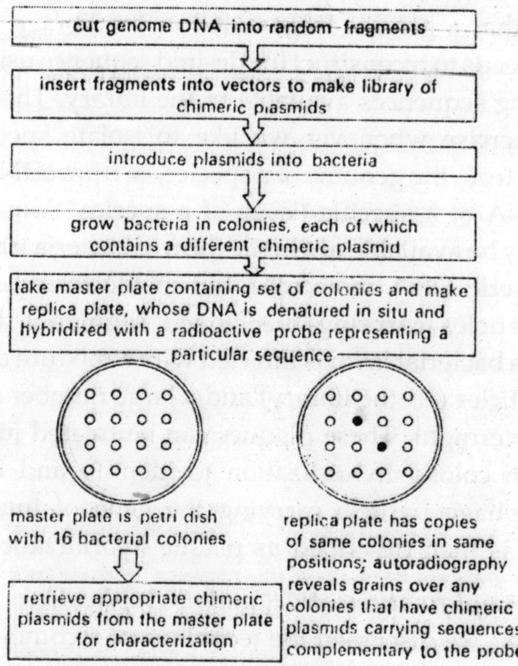

Colony hybridization technique for selection and isolation of a sequence complementary to a radioactively labeled probe.

Colony (or plaque) Hybridization for Screening of Libraries : Once a genomic library or cDNA library is available, we may like to use it for isolation of a gene sequence. This can be achieved by colony hybridization technique illustrated in Figure. In this technique, bacteria carrying chimeric vectors are grown into colonies, which are lysed on nitrocellulose filters. Their DNA is denatured *in situ* and fixed on the filter, which is hybridized with a radioactively labeled probe carrying a sequence related to the gene to be isolated (usually a cloned cDNA for screening of a genomic library). Colonies carrying this sequence will be identified by dark spots after autoradiography, so that the original chimeric vector carrying the desired gene sequence can be recovered from the original master plate and used for further experiments.

It is possible that a probe may identify more than one clones or that a gene is fragmented in the library. In such a case, one needs to reconstruct the desired sequence using several overlapping sequences available in the library. This is a very routine exercise whenever we like to isolate specific DNA sequences from the genome of a species, or from cDNA derived from mRNA of a specific tissue of a species. Sometimes the library may be available not in the form of bacteria transformed with chimeric DNA molecules, but in the form of chimeric phage particles carrying the cloned segments. In such a situation, a bacterial lawn is infected with a mixture of chimeric phage particles (*i.e.* the library) and a large number of plaques develop overnight. These plaques can be treated just like the colonies in colony hybridization to identify and isolate the chimeric phage particle carrying the gene of interest. This technique is then described as plaque hybridization.

Gene Amplification—PCR and its Applications : Earlier in this chapter, we discussed the technique of recombinant DNA and gene cloning. This technique permits us to obtain an unlimited supply of identical copies of a gene sequence or

DNA segment, which is cloned in a prokaryotic or a eukaryotic cell with the help of a vector. The technique (discovered around 1975), proved extremely useful in all experiments of molecular biology and genetic engineering and thus became an essential tool in all molecular biology laboratories. In 1985, yet another remarkable tool in molecular biology was discovered (by Kary Mullis), which is known as 'polymerase chain reaction', nick named as 'peoples choice reaction'. The methodology is so important that the prestigious journal *Science* published from USA considered PCR as the major scientific development of the year 1989, and had chosen Taq_1 DNA polymerase, the enzyme used in PCR, as the molecule of the year 1989. The polymerase chain reaction is such a powerful technique that it may replace completely, the gene cloning with vectors in due course of time.

Basic Polymerase Chain Reaction (PCR)

In order to understand PCR, the readers should be familiar with the mechanism of DNA replication within the cells. The DNA replication involves polymerization of nucleotides using a template DNA strand with the help of an enzyme 'DNA polymerase' but this reaction invariably requires a primer strand to which further nucleotides can be added using the DNA polymerase enzyme. If the primer strand is not available the reaction can not proceed. In the living cells this primer strand is not a DNA strand but is a small single stranded RNA molecule synthesized with the help of a primase or an RNA polymerase enzyme. In PCR, a similar reaction takes place in an Eppendorf tube, where the primer strand is added from outside in the form of a deoxyoligonucleotide and DNA polymerase enzyme is added to help in polymerization. Unlimited supply of amplified DNA is obtained by repeating the reaction, which is made possible by regular denaturation of freshly synthesized double stranded DNA molecules by heating it to 90-98°C. At this high temperature the two strands separate.

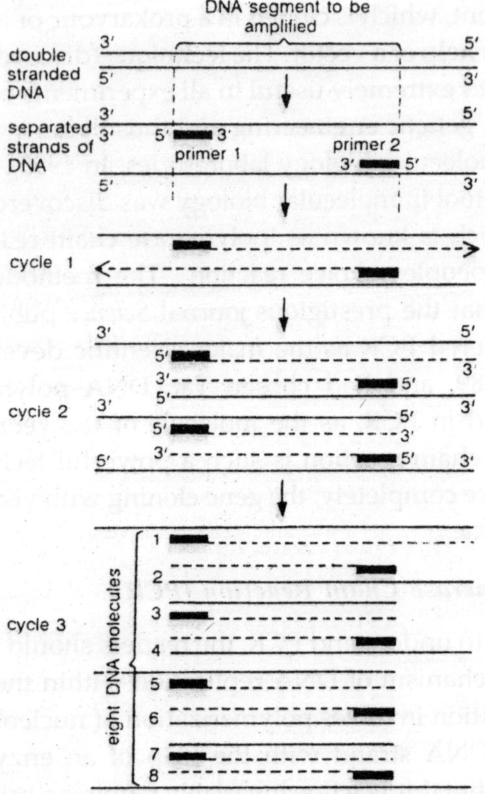

Steps involved in basic polymerase chain reaction or PCR.

When discovered initially, the PCR involved the following steps. The denatured DNA (at 90-98°C) and two primers, one recognizing each strand and bordering the sequence to be amplified were mixed. The mixture was then cooled to 40-60°C. This allowed the primers to find their complementary strands through renaturation. To this mixtuie was then added (i) DNA polymerase enzyme and (ii) all the four essential nucleoside triphosphates.

The mixture was incubated and this allowed synthesis of complementary strands using primers with the help of DNA polymerase enzyme in the usual manner. The process was repeated 20-30 times, so that in a single afternoon, a billion copies of the sequence flanked by the left and right primers could be produced (Fig.). In order to continue the synthesis, the temperature of the mixture was alternately increased (for denaturation) and decreased (for renaturation) and therefore, during temperature rise the enzymatic activity of DNA polymerase would be destroyed and one may have to add fresh aliquot of enzyme in each cycle of amplification. This problem could be overcome by the discovery of 'Taq polymerase', the DNA polymerase isolated from *Thermus aquaticus* growing in hot springs. The enzyme acts best at 72°C and the denaturation temperature of 90°C does not destroy its enzymatic activity. This enzyme allowed the automation of the entire process and now automatic PCR thermal cyclers are available for a price of around Rs. 2.0 to 4.0 lakhs, which can amplify DNA sequences at a fast speed unattended.

As pointed out earlier, due to its ease, PCR may eventually replace gene cloning technique. Also while in gene cloning, starting material is needed in microgram quantities, in PCR, DNA in rianogram quantities (1/1000th of microgram = 10^{-9}g) is enough. A comparison between PCR and gene cloning is given in Table.

Comparison of PCR and Gene Cloning

Parameter	*PCR*	*Gene cloning*
1. Final result	selective amplification of specific sequence	selective amplification of specific sequence
2. Manipulation	*in vitro*	*in vitro* & *in vivo*
3. Selectivity of specific segment from complex DNA	first step	last step
4. Concentration of starting material	nanogram (ng)	microgram (mg)

Contd...

Parameter	*PCR*	*Gene cloning*
5. Biological reagents required	DNA	restriction enzyme, ligase, vector, bacterial cell
6. Automation	yes	no
7. Labour intensive	no	yes
8. Error probability	less	more
9. Applications	more	less
10. Cost	less	more
11.User's skill	not required	required
12.Time for a typical experiment	four hours	two to four days

Different Schemes of PCR

We discussed the basic outline of PCR technique, outlining how billions of copies of a DNA sequence can be obtained within a few hours, without the use of a vector and a host as done in gene cloning. However, there are several variations to this basic PCR technique, which allows varied applications of this technique.

Inverse PCR: In this technique, amplification of those DNA sequences takes place, which are away from the primers and not those which are flanked by the primers. For instance if the border sequences of a DNA segment are not known and those of a vector are known, then the sequence to be amplified may be cloned in the vector and border sequences of vector may be used as primers in such a way that the polymerization proceeds in inverse direction *i.e.* towards the inserted segment, and not away from it towards the DNA sequence of vector from which primers have been derived (Fig.a). Similarly, if the gene sequence is known, one can use its border sequences as primers to amplify the sequences flanking this gene *e.g.* the regulatory sequences (Fig. b).

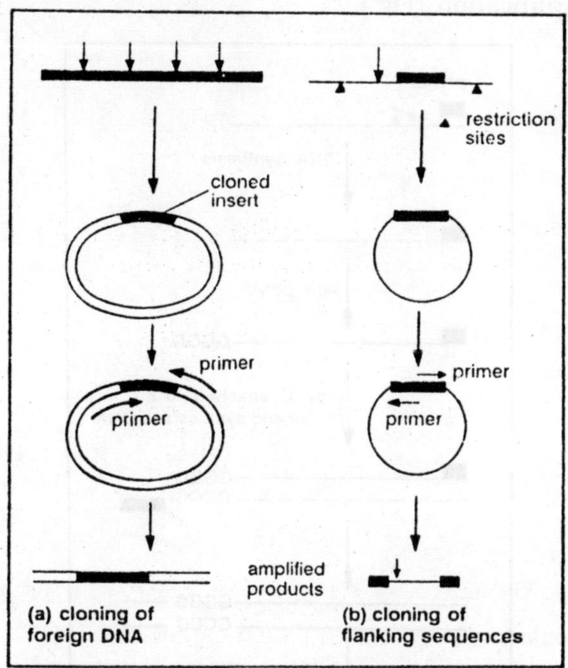

Inverse PCR : (a) use of vector border sequences as primers for amplification of inserted sequence, (b) use of border sequences of a gene to amplify its flanking regulatory sequences.

Anchored PCR: In the basic PCR technique and the inverse PCR, one has to use two primers representing the sequences lying at the ends of sequences to be amplified. But sometimes, we may have knowledge about the sequence at only one of the two ends of the DNA sequence to be amplified. In such cases anchored PCR may be used, which will utilize only one primer instead of two primers. In this technique, due to the use of one primer, only one strand will be copied first, after which a poly G will be attached at the end of the newly synthesized strand. This newly synthesized strand with poly G tail at its 3' end will then become template for the daughter strand synthesis utilizing an anchor primer with which a poly C sequence is linked to complement with poly G of the template. In the next cycle, both the original primer and anchored primer will be used for gene amplification (Fig.).

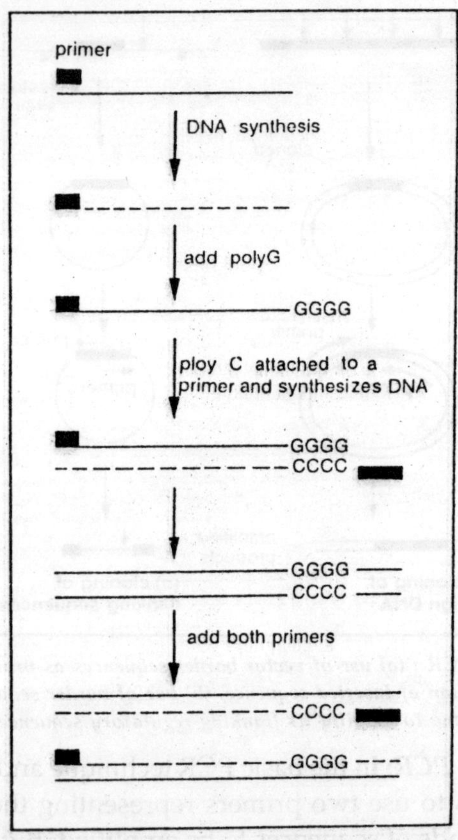

Anchored PCR, using only one primer at the initial stage of amplification.

PCR for Site Directed Mutagenesis: This technique is used for introducing mutations at the desired place in a DNA sequence by altering the sequences of primers (Fig.). Since mutations are introduced only through primers, mutations are limited to the ends of the gene sequence. A variation of this technique allows mutations at any place of interest in the gene. The method is described as *overlap extension* which works as follows : (i) In two separate PCR reactions, a particular gene is amplified into two separate segments. In both reactions there is one primer at the end of the gene and the other internal to the sequence. The internal primers in two reactions are complementary to one another, so that the amplified products will have their ends internal to the original sequence. These internal ends of products in two reactions will overlap. The sequences of internal primers can be suitably modified to introduce alterations in the overlap region. (ii) The two PCR products are denatured and annealed, so that the internal ends in the overlap region will work as primers for each other. Extension of these primers results in the formation of a complex gene, with the mutation incorporated into it (Fig.).

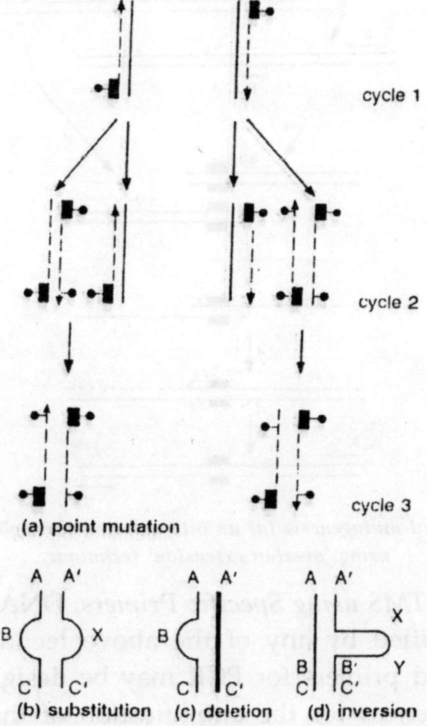

Directed mutagenesis (at the end of a gene sequence) through PCR, using a primer with altered base sequence.

RAPDs using Random Primers: PCR has also been used for developing molecular markers termed 'Randomly Amplified Polymorphic DNAs' (RAPDs, pronounced as 'rapids'). In this technique instead of using a pair of primers, a single short oligonucleotide primer is used, which binds to many different loci and allows amplification of random sequences from a complex DNA template (*e.g.* whole genomic DNA from a plant or animal). This has several advantages over RFLPs earlier used as molecular markers.

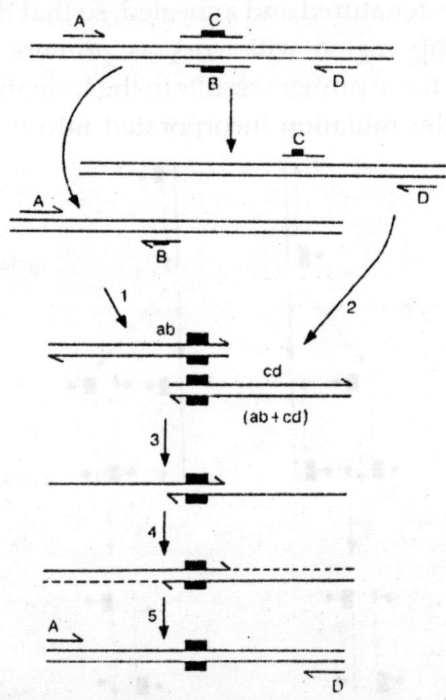

Directed mutagenesis (at an internal site) through PCR, using 'overlap extension' technique.

STS and STMS using Specific Primers: DNA fragments of interest identified by any of the above technique may be sequenced and primers for PCR may be designed from the flanking sequences for the amplification of the intervening sequence. When a fragment used for RFLP is used in this manner, then sites are described as Sequence Tagged Sites

(STS) and when microsatellites or simple sequence repeats (SSRs) are used, the characterized sites are described as Sequence Tagged Microsatellite Sites (STMSs). Both STS and STMS loci can be amplified on PCR using specific primers as above and the products can be visualized on agarose or PAGE (polyacrylamide gel electropheresis).

DNA Chip Technology and Microarrays

During mid 1990s, DNA chips with microarrays of DNA samples became available to achieve very fast speeds in generating information about DNA sequences. This is considered a major technological advance in the field of molecular biology and is compared with the role of semiconductors in the field of electronics. These microarrays are prepared on solid surfaces (glass plates/slides are often used) and represent high-density miniaturized arrays of molecular samples, thus facilitating the screening of genomic DNA or cDNA samples for the presence of 1 in 100,000 or even more DNA sequences in a single hybridization.

The sequences on DNA chips may be oligonucleotides (20-mers) of known sequences or these may be cDNA sequences with known function. These microarays are hybridized with an unknown labelled DNA sample, and the hybridization patterns are analysed with computer devices. The major uses of these microarrays include the following : (i) DNA sequencing by hybridization (SBH), leading to the detection of single nucleotide polymorphisms (SNPs), as has been done for human genome, (ii) diagnostics and genetic mapping as has been done for CFTR (cystic fibrosis) and BRCA1 (cancer susceptibility) in humans, (iii) gene expression studies and (iv) proteomics involving study of protein - protein interaction. It is expected that these DNA chips (microarrays) will be extensively utilized in the field of molecular biology, in the 21st century for making major advances at a very fast speed.

9

Transfer Methods

In recent years (1980s and 1990s), the use of novel methods for transfer of genes over a wide range of organisms, leading to the production of transgenic animals and plants, has been recognized as one of the thrust areas of biotechnology. Transfer of genes by conventional methods of animal and plant breeding, is possible over a very narrow range *i.e.* within a species.

Even by non-conventional methods involving tissue or organ culture, genes can be transferred only at the interspecific or intergeneric level. The modern techniques of gene transfer using recombinant DNA technology, however, allow transfer of genes from even bacteria and insects (or even artificially synthesized genes) to higher plants and animals, thus increasing manifold the possibilities of improving animals (*e.g.* livestock) and plants.

The gene transfer methods, their successful use in production of a variety of desirable animals and plants, and the future possibilities of the exploitation of this technology for livestock and crop improvement will be briefly discussed in this chapter.

Gene Transfer Methods

Gene transfer methods, utilizing recombinant DNA technology, differ for animals and plants; while in animals, DNA is injected into either the pronuclei of embryonic cells (or fertilized egg) or the embryo stem (ES) cells, in plants, cells, protoplasts or meristematic embryonic tissues are used. This basic difference is due to the availability of the techniques of *in vitro* fertilization (IVF) and embryo transfer (ET) in animals and those of protoplast culture and regeneration of protoplasts in whole plants in higher plants. IVF and ET is not successful in plants (instead embryo culture is possible). Similarly cell culture and regeneration of a cell into a whole animal is not possible in higher animals. In view of this, gene transfer methods for animals and plants will be discussed separately.

Gene Transfer (Transfection) Methods in Animals : In bacteria and other microbes, or even in higher plants, the uptake of genes by cells is often described as 'transformation'. In animals, this term is replaced by the term 'transfection'

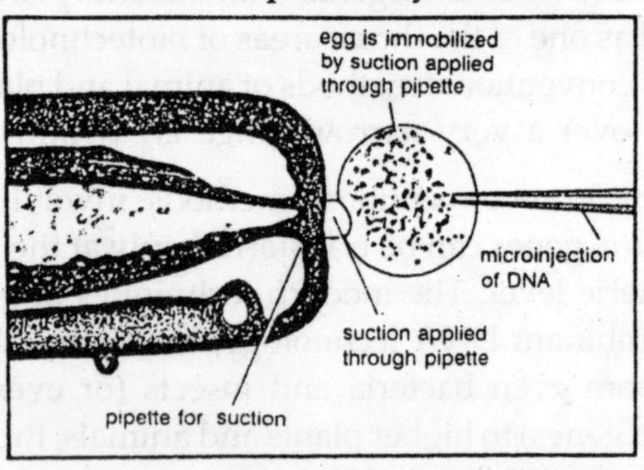

Microinjection of foreign DNA into a fertilized egg for transfer of a gene leading to the production of transgenic animals.

because the term transformation is used in animals to describe phenotypic alteration of cells. Transfection or gene transfer in animals can be carried out at the cellular level and the transfected cells may be used for a variety of purposes including the following :

(i) production of chemicals and pharmaceutical drugs;

(ii) study of structure and function of genes and

(iii) production of transgenic animals for increased milk production, increased growth rate of livestock and fish, etc., large scale production of valuable proteins (to be extracted from milk, urine and blood ; this is described as 'molecular farming') and for improvement of wool in sheep.

The transfection methods may employ the fertilized eggs/ embryos or the cultured cells. The technique and the purpose of transfection may differ in the two cases.

Transfer of Genes to Fertilized Eggs or Embryos : The transfection of fertilized egg may involve transfer of whole nuclei, whole chromosomes (or fragments) or DNA segments.

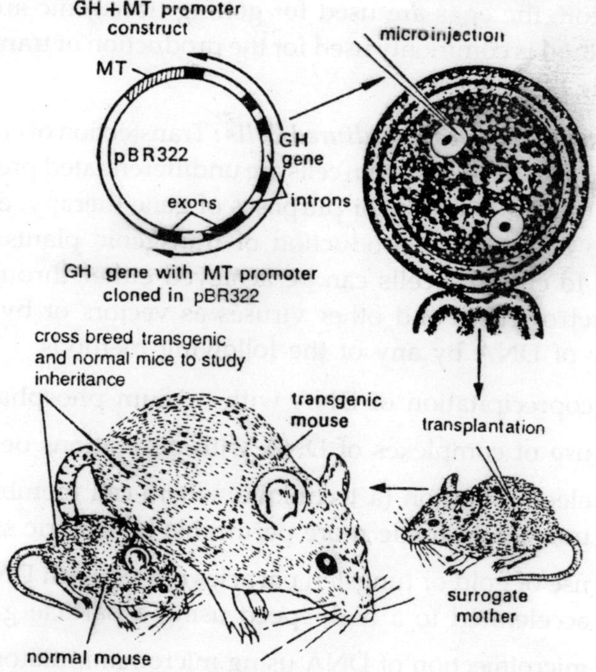

Different steps in the production of transgenic mice through microinjection of eggs.

(i) For the transfer of whole nuclei, the egg cells are treated with cytochalasin-B and subjected to centrifugation causing enucleation (nucleus is removed). Incubation of desirable karyoplasts with these enucleated eggs leads to the transfer of whole nuclei in presence of polyethylene glycol (PEG); during early 1997 this technique was reported to have been utilized for transfer of a somatic nucleus to an egg, leading to the production of a clone of sheep named Dolly; (ii) For transfer of whole chromosomes, the chromosomes are first isolated from metaphase cells by hypotonic lysis and may be fractionated using density centrifugation or flow cytophotometry. Individual chromosomes or fragments thus isolated are then incubated with whole cells (eggs) for incorporation of chromosomes into nuclei; (iii) Microinjection of DNA fragments into fertilized eggs is the most commonly used method (Fig.). Several hundred copies of the DNA fragment are utilized and after stable integration, the eggs are used for getting transgenic animals. This method is commonly used for the production of transgenic mice (Fig.).

Transfer of Genes to Cultured Cells : Transfection of cultured cells (usually stem cells; stem cells are undifferentiated precursor cells) is often attempted for purposes of gene therapy, etc. but can also be used for production of transgenic plants. Gene transfer to cultured cells can be achieved either through the use of retroviruses and other viruses as vectors or by direct delivery of DNA by any of the following methods:

(i) coprecipitation of DNA with calcium phosphate;

(ii) use of complexes of DNA with polycations or lipids;

(iii) electroporation (a technique, where cell membrane is transiently made more permeable by electric shocks);

(iv) use of gold or tungsten particles coated with DNA and accelerated to a high speed using a particle gun;

(v) microinjection of DNA using micromanipulator (Fig.); and

(vi) use of liposomes or sphaeroplasts, which are vesicles carrying DNA. In future more efficient methods of gene transfer to cultured cells in animals will become available.

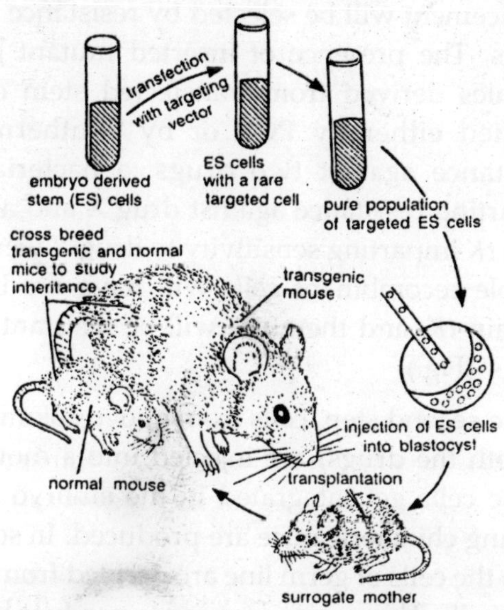

Transfection of stem cells and injection of derived stem cells into blastocyst, leading to the production of transgenic mice.

Selective Gene knockout by Homologous Recombination in Mouse: We know, that during the production of transgenic animals or plants, the transgene is inserted into the chromosome at random. However, once in a thousand cases, it may replace one of the two homologous normal genes in the recipient, through homologous recombination. These rare 'gene targeting' events can be used for direct gene replacement thus inactivating a specific gene.

The technique has also been used to produce a mouse with a missing gene and is described as 'gene knockout'. This technique of gene knockout is being utilized widely to dissect the functions of specific mammalian genes. The gene knockout technique involves the following two steps:

(i) In the first step a desired mutant gene is inserted in embryonic stem cells (ES cells) using a vector. The cloned mutant gene is so designed that only the rare homologous recombinants resulting by gene replacement will be selected by resistance against two drugs. The presence of inserted mutant gene in cell colonies derived from transfected stem cells can be verified either by PCR or by Southern blots. For resistance against two drugs, a bacterial gene *neo* imparting resistance against drug X and another viral gene *tK* imparting sensitivity to drug Y were used. The double recombinant cells will have *neo* but will not contain *tK*, and therefore will be resistant to both the drugs (Fig.),

(ii) In the second step, cells from above colonies (resistant to both the drugs) are injected into a mouse embryo. These cells get integrated in the embryo and normal looking chimaeric mice are produced. In some of these mice the cells of germ line are derived from transfected stem cells. These mice are used to produce heterozygote male and female mice carrying the transgene. When these male and female heterozygotes, are mated, one fourth of the progeny will be homozygous for the presence of altered gene and for the loss of the normal gene.

Gene Transfer Methods in Plants : Gene transfer methods in plants make use of a variety of target cell types, which include protoplasts or cultured cells, meristem cells from immature embryos, shoots or flowers and pollen or zygotes. The methods can be broadly classified into two groups :

(i) *Agrobacterium* Ti or Ri plasmid mediated gene transfer and

(ii) physical delivery or DNA mediated gene transfer (DMGT).

Transfer of genes using Agrobacterium Ti and Ri plasmids: Ti plasmid of *Agrobacterium tumefaciens* (causing 'crown gall' disease; Fig.) and Ri plasmid of *A. rhizogenes* (causing 'hairy root' disease) have been effectively used as vectors for gene transfer to plant cells. The Ti plasmid (and also Ri Plasmid) has a region called vir region (having six operons *vir A, B, C, D, E, G)* responsible for virulence towards host and T-DNA region, which is transferred to the host, along with the gene intended to be transferred.

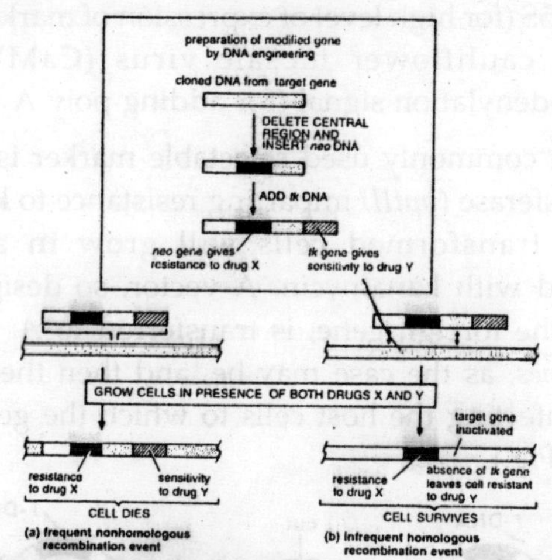

Steps involved in the selective knockout by homologous recombination in the mouse figure redrawn from Alberts et at., 'Molecular Biology of the Cell', 1994).

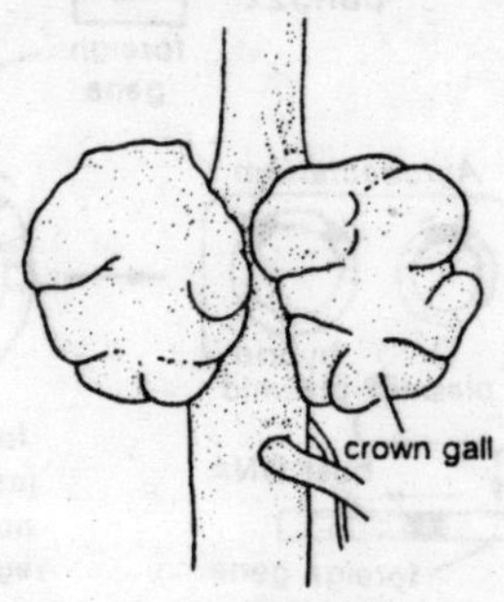

Tumour formation and leafy structure, on the stem of a plant, induced by Agrobacterium tumefaciens.

Vectors based on Ti and Ri plasmids have been designed with the following properties :

(i) they do not cause the disease, since they are disarmed by removing genes causing the disease;

(ii) they carry sites for insertion of foreign gene intended to be transferred and

(iii) they carry selectable markers (genes which will help in selecting the transformed cells), a powerful promoter like 35S (for high level of expression of marker) derived from cauliflower mosaic virus (CaMV), and a polyadenylation signal (for adding poly A to mRNA).

The most commonly used selectable marker is neomycin phosphotransferase *(nptII)* imparting resistance to kanamycin, so that the transformed cells will grow in a medium supplemented with kanamycin. A vector, so designed, after insertion of the foreign gene, is transferred to *A. tumefaciens* or *A. rhizogenes*, as the case may be, and then the bacterium is used for infecting the host cells to which the gene is to be transferred (Fig.).

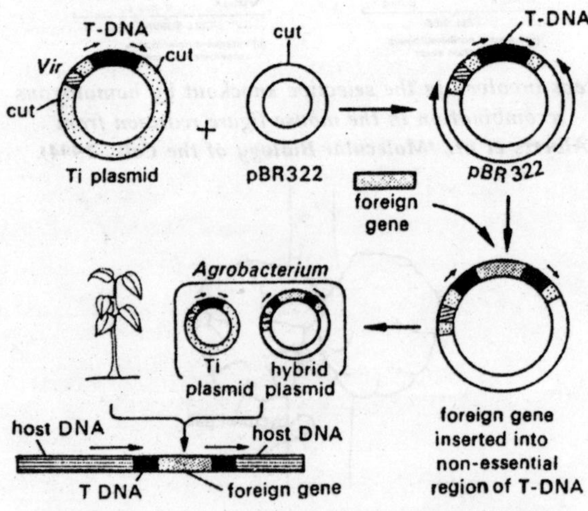

Different steps for the transfer of a foreign gene into a tobacco plant using A. tumefaciens.

For gene transfer, the explants commonly used include protoplasts, suspension cultured cells, callus cell clumps, thin cell layers (epidermis), tissue slices, leaf discs, and sections of roots, stems or floral tissues. These explants are incubated or cocultivated with *Agrobacterium* carrying the vector with the desired foreign gene. The transformed cells or colonies are selected due to the selectable marker and used for regeneration of whole plants which are expected to be transgenic in nature. These regenerated plants can again be tested for the transfer of gene, with the help of one of the several scoreable or screenable markers (also called reporter genes), which is also transferred along with the desirable gene. The most commonly used reporter genes are *nptll* (neomycin phosphotransferase II gene), *cat* (chloram-phenicol acetyltransferase gene), *gus* (β glucuronidase gene) and *lux* (luciferase gene).

Transfer of Genes using Physical Delivery Methods: Although *Agrobacterium* mediated gene transfer was extensively utilized for production of transgenic dicotyledonous plants; initially it was not successful in monocotyledonous plants, since *Agrobacterium* generally does not infect them. Therefore, for cereals and also for some dicots (*e.g.* legumes), where regeneration was difficult, initially the transgenic plants could not be produced using *Agrobacterium* (more recently, it has been possible to achieve *Agrobacterium* mediated transformation in monocots including cereals like maize, rice and wheat). In view of this, DNA mediated gene transfer (DMGT) was successfully tried in a wide range of plants including both monocotyledons and dicotyledons. These DMGT methods include the following :

(i) PEG (polyethylene glycol) stimulated DNA uptake involves the use of 15-25% PEG, which stimulates uptake of DNA by endocytosis without any gross damage to protoplasts. The transformed protoplasts can be selected on a selection medium.

(ii) Microinjection involves injection of DNA, using micropipettes with 0.5-10 μm diameter tip. In this technique, the recipient cells are immobilized on a solid support (cover slip or slide, etc.) or artificially bound or held by a pipette under suction (Fig.) Specially designed micromanipulators are used for this purpose.

(iii) Macroinjection of DNA, employing needles with diameters greater than cell diameter, has been tried in rye *(Secale cereale)*. DNA was injected into the stem below the immature floral meristem, so as to reach the sporogenous tissue, leading to the production of transgenic rye plants. Unfortunately this technique was not successful when tried with other cereals, making the validity of earlier experiments doubtful.

(iv) Microprojectiles (gold or tungsten particles 'coated with DNA' 1-3 μm in diameter) are carried by a 'macroprojectile' (or bullet), and are accelerated into target cells, where they penetrate the cell wall, leaving DNA to be incorporated (Fig.). This technique has been successfully used for transformation in soybean, tobacco, maize, rice, wheat, oats and tall fescue. This method is now believed to be universal in its application and has been extensively utilized in 1990s for gene transfer in plants, irrespective of species and genotype.

(v) Electroporation is also used for DNA uptake into the protoplasts, but may not be suitable, if the target cells are available in the form of multicellular explants with cell walls.

(vi) Liposome mediated gene transfer involves encasement of DNA in lipid bags, which can be made to fuse with protoplasts with the help of a peg. This technique has been used successfully for gene transfer in a number of plant species.

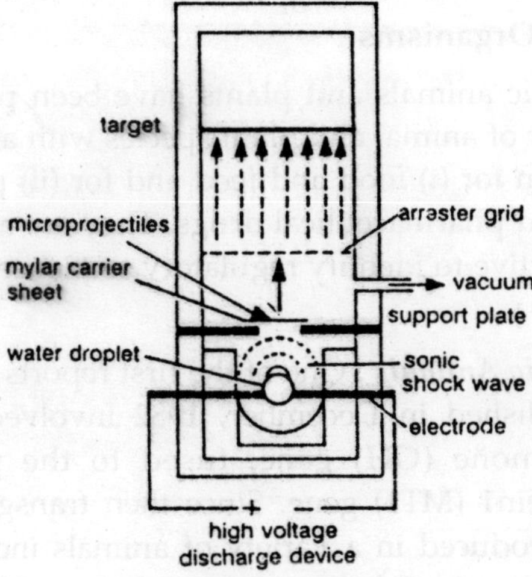

(a) electrostatic device

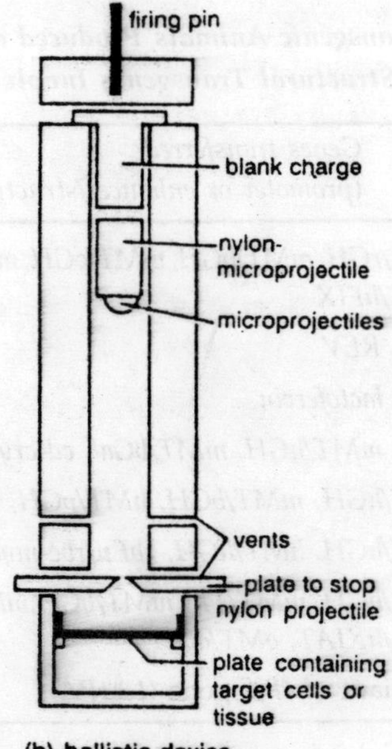

(b) ballistic device

Particle guns for accelerating microprojectiles carrying a gene of interest to the target cells (for transfer of gene).

Transgenic Organisms

Transgenic animals and plants have been produced in a large number of animal and plant species with an objective to improve them for (i) food and feed and for (ii) production of chemicals and pharmaceutical drugs. They are also produced with an objective to identify regulatory sequences of a variety of genes.

Transgenic Animals : One of the first reports of transgenic animals published in December, 1982 involved transfer of growth hormone (GH) gene, fused to the promoter of metallothionein1 (MT1) gene. Since then transgenic animals have been produced in a variety of animals including mice, rabbits, sheep, pigs, goats, cows, fish, etc. (Table). These transgenic animals will be utilized for their (i) efficiency in utilizing feed; (ii) ability to give leaner meat; (iii) ability to

A List of Transgenic Animals Produced and the Promoter, Enhancer and Structural Transgenes Involved in each Case

	Animal	*Genes transferred (promoter or enhancer/structural genes)*
1.	Mouse	*mMT/rGH, mMT/bGH, mMT/oGH, mMT/hGH, mMT/hGRF, inMT/hFIX*
2.	Chicken	*ALV, REV*
3.	Cow	*BPV, lactoferrin*
4.	Fish	*hGH, mMT/hGH, mMT/bGal, cd-crystallin SV/hygro, AFP.*
5.	Pig	*mMT/hGH, mMT/bGH, hMT/pGH, MLV/rGH, bPRLbGH*
6.	Rabbit	*mMT/hGH. hMT/hGH, rbEu/rbc-myc*
7.	Sheep	*mMT/hGH, mMT/TK. mMT/bGH,mMT/hGRF, oBLG/hFIX, oBLG/hXIAT, oMT/oGH*
8.	Goats	a variant of *tPA* gene *(LAtPA)*

Note : The list of genes; promoter/or enhancer is separated from structural gene by a slash (/). The source of the gene is given by a lower case preceding the gene : *b* = bovine; c = chicken; h = human; *m* = mouse; *o* = ovine; *p* = porcine; *r* = rat; rb = rabbit.

Gene abbreviations—ALV = avian leukosis virus; *xlAT* = xl anti-trypsin; *BPV* = bovine papilloma virus; *Eu* = immunoglobulin heavy chain; *FIX* = factor IX; *GH* = growth hormone; *GRF* = growth releasing factor; *Gal* = *b* galactosidase; *hygro* = hygromycin; *BLG* = B-lactoglobulin; *MT* = metallothionein; *MLV* = moloney murine leukemia virus; c-*myc* = murine leukemia virus; *REV* = reticuloendotheliosis; *PRL* = prolactin; *SV* = SV40; *TK* = thymidine kinase; *AFP* = antifreeze; t*PA* = human tissue-type plasminogen activator.

grow to marketable size sooner; (iv) resistance to certain diseases. They are also proposed to be used as bioreactors, so that they will secrete valuable recombinant proteins and Pharmaceuticals in their milk, blood and urine, which can be used for extraction of these drugs. This is described as 'molecular farming' or 'molecular pharming'.

Transgenic Plants

During the last 15 years, transgenic plants have been produced in as many as 70 different species (Table), and transgenic varieties belonging to atleast eight species have already been released for commercial cultivation. The first field trials of a herbicide resistant transgenic tobacco were conducted in USA in 1986. Ever since, during 1986-95, more than 3500 field trials of atleast 56 transgenic crop varieties (representing only 8 species) were conducted at more than 1500 sites in 34 countries. Surprisingly, China was the first country to grow transgenics for commercial purposes in 1993-94 for seed multiplication. This was followed by several countries including USA, Australia and a number of countries in Europe.

According to some estimates, at the global level, the land occupied by transgenic crops increased from a meagre 7 million acres in 1996 to approximately 125 million acres in the year 2001. These crops included herbicide tolerant and/or insect resistant corn, cotton, soyabean potato and tomato, virus resistant papaya and potato, high oleic acid canola and soyabean and delayed ripening tomato. The global market for transgenic crops is estimated to reach 6 billion US dollars in the year 2005.

Transgenic plants have now been produced in all major crops, particularly for the improvement of those agronomic traits (controlled by one or more major genes) for which genetic variability is not available within the corresponding gene pool (to be utilized in conventional plant breeding). Some of the traits improved through transgenic approach include the following (these traits also include some difficult traits that required manipulation of biosynthetic pathways):

(i) resistance against biotic stresses like insects, bacterial and fungal pathogens, viruses and weeds; recently transgenic plants have also been produced that are resistant not to one, but to several viruses or to several other disease causing pathogens (in this case using 'master-switch' gene, NPRI regulating many pathogen related or PR genes);

(ii) resistance against abiotic stresses including cold, drought, heat, acidity, metal (*e.g.*, AI, Hg) toxicity, etc.

(iii) improvement in harvest index using phytochrome genes;

A List of Some Higher Plants where Transgenic Plants have been Produced, Using different Methods.

I. Herbaceous Dicotyledons

1. *Nicotiana tabacum* (tobacco)
2. *N. plumbaginifolia* (wild tobacco)
3. *Petunia hybrida* (petunia)
4. *Lycopersicon esculentum* (tomato)
5. *Solanum tuberosum* (potato)
6. *Solanum melongena* (eggplant)

7. *Arabidopsis thaliana*
8. *Lactuca saliva* (lettuce)
9. *Apium graveolens* (celery)
10. *Helianthus annuus* (sunflower)
11. *Linum usitatissimun* (flax)
12. Br*assica napus* (oilseed rape; canola)
13. *Brassica oleracea* (cauliflower)
14. *Brassica oleracea* var. capitata (cabbage)
15. *Brassica rapa* syn. B. *campestris*
16. *Gossvpium hirsulum* (cotton)
17. *Beta vulgaris* (sugarbeet)
18. *Glycine inax* (soybean)
19. *Pisum sativum* (pea)
20. *Medicago sativa* (alfalfa)
21. *M. varia*
22. *Lotus corniculatusm* (lotus)
23. *Vigmt aconitifolia*
24. *Cucumis sativas* (cucumber)
25. *Cucumis melo* (muskmelon)
26. *Cichorium intybus* (chicory)
27. *Daucus carota* (carrot)
28. *Armoracia* sp. (horse radish)
29. *Glycorrhiza glabra* (licorice)
30. *Digitalis purpurea* (foxglove)
31. *Ipomoca batatas* (sweet potato)
32. *Ipomoca purpurea* (morning glory)

33. *Fragaria* sp. (strawberry)
34. *Actinidia* (kiwi)
35. *Carica papaya* (papaya)
36. *Vitis vinifera* (grape)
37. *Vaccinium macrocarpon* (cranberry)
38. *Dianthus caryophyllus* (carnation)
39. *Chrysanthemum* sp. (chrysanthemum)
40. *Rosa* sp. (rose)

II. Woody Dicotyledons

41. *Populus* sp. (poplar)
42. *Malus sylvestris* (apple)
43. *Pyrus communis* (pear)
44. *Azadirachta indica* (neem)
45. *Juglans regia* (walnut)

III. Monocotyledons

46. *Asparagus* sp. *(asparagus)*
47. *Dactylis glomerata* (orchard grass)
48. *Secale cereale* (rye)
49. *Oryza sativa* (rice)
50. *Zea mays* (corn)
51. *Triticum aestivum* (wheat)
52. *Avena saliva* (oats)
53. *Festuca arundinacea* (tall fescue)

IV. Gymnosperms (a conifer)

54. *Picea glauca* (white spruce)

(iv) improvement in biomass production using hemoglobin gene from bacteria;

(v) improvement in vitamin A content;

(vi) improvement in iron content in crops like rice;

(vii) improvement in nutritional quality of cereals and legumes, through alteration in the protein content and composition (*e.g.* increase in lysine content);

(viii) improvement in the content and composition (fatty acid composition) of edible oils in rapeseed and other oil-seed crops;

(ix) production of transgenics that can be used as edible vaccines;

(x) production of transgenic with male sterility fertility-restoration system to be used for hybrid seed production;

(xi) development of caffeine free transgenic coffee;

(xii) improvement in phosphorus uptake through the use of phosphorus transporter genes, to reduce dependence on fertilizer application;

(xiii) bruise resistance and delayed ripening in tomato, to make them suitable for transport and longer shelf life;

(xiv) improvement in starch content and composition (relative proportions of amylose and amylopectin) in crops like potato;

(xv) development of seedless eggplant and high solid, seedless tomato (due to parthenocarpy), thus reducing the dependence of fruit production on pollination and making them suitable for food processing (in tomato fruits, seeds need to be removed and water content needs to be reduced for making ketchup/sauce, etc.);

(xvi) improvement in architecture, colour, fragrance and vase life of flowers of commercial value.' Transgenic plants for some of the above traits will be described briefly.

Transgenic Plants with Herbicide Resistance: Due to increasing concern about contamination of environment due to herbicides, new herbicides are being developed that are safer and biodegradable. This has necessitated the development of resistance in crop plants against these new and safer herbicides. These newer herbicides affect processes like photosynthesis or biosynthesis of essential amino acids (Table). Transgenic plants resistant to these herbicides have been produced, utilizing one of the following two approaches:

(i) Either the target protein is overexpressed (*e.g.* EPSPS) or it should become insensitive to the herbicide (*e.g.* ALS).

(ii) A pathway is introduced, which will detoxify the herbicide. For instance, glutathione-S transferase (GST) detoxifies atrazine, 'nitrilase' coded by gene *bxn*, detoxifies 'bromoxynil' and 'phosphinothricin acetyl transferase (PAT)' coded by *bar* gene, detoxifies 'L-phosphinothricin (PPT)'. (For more details, see Table).

Transgenic Plants with Insect Resistance : Genes for insect resistance from three different sources are being transferred for developing insect resistance (i) *bt2* gene encoding Bt toxin. derived from *Bacillus thuringiensis;* (ii) cowpea trypsin inhibitor gene (CpTi) from cowpea *(Vigna\ unguiculata)* and (iii) genes for other insecticidal secondary metabolites derived from a number of legumes. In all these cases, successful attempts to transfer genes for insect resistance have already been made. Insect resistant transgenic plants derived thus are being field tested and in some cases released for commercial cultivation. Transgenic plants with resistance against viruses. Three different approaches were used for developing virus resistant transgenic plants :

Mechanisms of Action of the different Herbicides and Basis of Achieving Resistance against them in Transgenic Plants

Active principle of herbicide	Inhibited Pathway	Target product	Use	Basis of resistance
I. Amino Acid Biosynthesis Inhibitors				
1. Glyphosate	Aromatic amino acid biosynthesis	EPSPS	Broad spectrum	Overexpression of *EPSPS* gene; bacterial *aroA* gene
2. Sulphonylurea and Imidazolinones	Branched chain amino acids	ALS	Selected crops	Mutant *ALS* gene
3. Phosphinothricin	Glutamine biosynthesis	GS	Broad spectrum	Gene amplification, *bar* gene; detoxification
II. Photosynthesis Inhibitors				
4. Atrazine detoxification	Photosystem II	QB	Selected crops	Mutant *PsbA* gene; GST gene;
5. Bromoxynil	Photosynthesis	—	Selected crops	*bxn* gene; detoxification

(i) Genes for virus coat protein or capsid protein (CP) were transferred to tomato, alfalfa, tobacco, potato, melon and rice for developing resistance against a variety of viruses (*e.g.* TMV, PVX and PVY, etc.).

(ii) Gene for nucleocapsid protein (N) was transferred from any tospovirus like tomato spotted wilt virus (TSWV) and provided resistance against tospoviruses, which causes considerable damage to crops like tomato, tobacco, groundnut, pepper, etc.

(iii) DNA fragments coding for satellite RNAs (also called virusoids) associated with viruses (*e.g.* 'cucumber mosaic virus' or CMV and 'tobacco ringspot virus' or TobRV) were transferred and provided resistance against these and other viruses.

Transgenic plants with resistance against bacterial and fungal pathogens. Several examples are now available, where genes imparting resistance against bacterial and fungal pathogens have been successfully transferred to tomato (against *Pseudomonas, Alternaria* and *Rhizoctonia)* and potato (against *Phytophthora).* Transgenic tomato plants for hard skin and improved flavour. In tomato using antisense RNA technology, transgenic plants have been produced which are either 'bruise resistant' (suitable for transport and storage) or exhibit 'delayed ripening' giving more time for ripening on the plant, thus permitting more time for sugar accumulation and also giving higher shelf life. These are given the name 'Flavr Savr'.

Transgenic plants for the production of hybrid seeds. During 1990-92, genes *barnase* (encoding a ribonuclease) and *barstar* (encoding a ribonuclease inhibitor) derived from *Bacillus amyloliquefaciens* were separately transferred to *Brassica napus* to produce 'male sterile' and 'fertility restorer' plants. The gene constructs were prepared where *barnase/barstar* was fused with TA29 promoter from tobacco,' permitting expression only in the anther causing male sterility or fertility restoration. This system will prove to be of immense value for hybrid seed production in crop plants in future (see Table).

Transgenes that are Inducible (by a Stimulus) and/or Exhibit Tissue Specific Expression in Transgenic Plants

Gene product	Donor plant	Transgenic host	Inducer stimulus	Specific host tissue
1. *rbcS*	Pea	Tobacco, petunia	Light	Green tissues
rbcS	*N. plumbaginifolia*	*N. Plumbaginifotia*, petunia, tobacco	Light	Green tissues
rbcS	Soyabean	Petunia	Light	Green tissues
2. *Cab*	Pea, wheat	Tobacco	Light	Green tissues
Cab	*N. plumbaginifolia*	Tobacco	Light	Green tissues
Cab	Wheat	Tobacco	Circadian clock	Green tissues
3. *ST-LSI*	Potato	Tobacco	Light	Green tissues
4. Chalconne synthase	*Antirrhinum majus*	Tobacco	Light	Flower
5. *Hse*	Maize	Petunia -	Heat	All tissues
Hse	Soyabean	Tobacco	Heat	All tissues
6. Protease inhibitor	Potato	Tobacco	Wounding	All tissues
7. Patatin	Potato	Potato	Tuber growth	Tuber
8. Leghaemoglobin	Soyabean	*Lotus*	Rhizobium	Nodule

Contd...

Gene product	*Donor plant*	*Transgenic host*	*Inducer stimulus*	*Specific host tissue*
9. Nodulin	Soyabean	*Lotus*	Rhizobium	Nodule
10. β-phaseolin	French bean	Tobacco	Rhizobium	Seed
11. β-conglycin	Soyabean	Petunia, tobacco	Seed development	Seed
12. Lectin	Bean	Tobacco	Seed development	Seed
	Soyabean	Tobacco	Seed development	Seed + root
13. Glutnin	Wheat	Tobacco	Seed development	Seed
14. Hordein	Barley	Tobacco	Seed development	Seed
15. Legumin	Pea	N. *plumbaginifolia*	Seed development	Seed
16. EPSP synthase	Petunia	Tobacco	Herbicide	Flower pollen
17. Ribonuclease	1. *Aspergillus oryzae*	Tobacco	Anther development	Anthers
	2. *Brassica amyloliquefaciens*	Tobacco	Anther development	Anthers
18. Ribonuclese inhibitor	*Brassica amyloliquefaciens*	Tobacco —	Anther development	Fhloem cells

Transgenic plants for molecular farming. Transgenic plants are also proposed to be used as factories or bioreactors for manufacturing speciality chemicals or pharmaceuticals. Sugars, fatty acids, starch, cellulose, rubber, wax, etc. are obtained from plants and transgenic plants can be produced to increase their production. Following are some examples :

(i) Transgenic tobacco plants with increased level of mannitol were produced using a gene for mannitol dehydrogenase (this also gave resistance against salinity). (ii) Transgenic potato plants with increased level of cyclodextrins or CDs were produced using a bacterial gene for cyclodextrin glucosyltransferase or CGTase. (CDs are useful for pharmaceutical delivery systems, flavour and odour enhancement and for removing undesirable compounds like caffeine from food). (iii) Transgenic *Arabidopsis* plants for production of polyhydroxybutyrate or PHB (a biodegradable thermoplastic polymer) were produced using two genes, *phbB* and *phbC*. (iv) Transgenic potato and tobacco plants for production of human serum albumin or HSA (in leaf tissue) could be successfully produced.

Transgenic plants for study of regulated gene expression. Transgenic plants have also been produced for identifying, regulatory sequences involved in the differential expression of gene activity in time and space or due to a specific stimulus like light, temperature, wound, etc. For this purpose, different lengths of upstream flanking sequences of a structural gene are fused with their own structural gene or with other unrelated structural genes studied in the transgenic plants thus produced. This allowed the identification of regulatory elements *e.g.* light response elements (LRE), endosperm box, heat shock element (HSE), etc. For instance when LRE was fused with an unrelated gene, the latter started behaving like a photosynthetic gene (expression in leaf and light, not in roots and dark). A number of genes, thus examined, are listed in Table.

10

Cytoplasmic Inheritance

Chromosome theory of inheritance suggested that Mendelian factors or genes were located on chromosomes. This theory was proved to hold good through a variety of experimental evidences.

Since the chromosome complement in male and female gametes obtained from same individual would be similar, reciprocal crosses (♀A × ♂ B : ♀B × ♂A) should give same results. The only exception to this expectation, is sex linked inheritance. Differences in reciprocal crosses involving sex linked characters can be easily explained on the basis of transmission of sex chromosomes.

Since chromosomes divide in a very precise manner both during mitosis as well as during meiosis, it is easy to draw a parallelism between chromosomes and genes.

Cytoplasm, however, does not divide in such a precise manner during cell division. Female gamete usually contributes more cytoplasm to the zygote, although rarely male gamete also contributes sufficient cytoplasm to contribute to the plastid or mitochondrial genome of the zygote. Consequently, for characters having cytoplasmic control, differences in reciprocal crosses would be observed. Inheritance in these cases would

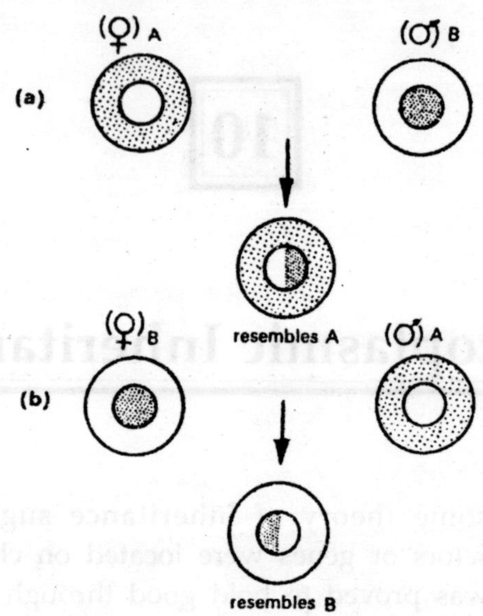

Maternal inheritance showing difference in reciprocal crosses, (a) ♀A × ♂B, (b) ♀B × ♂A.

be mainly of maternal type as shown in Figure. As can be seen in the figure, if two strains A and B respectively having genotypes *AA* and *BB* and cytoplasms a and b are crossed reciprocally, we will get two hybrids *AB(a)* and *AB(b)* (cytoplasm is indicated in parentheses). In case of maternal effect, *AB(a)* and *AB(b)*, despite having same nuclear genotype will differ. *AB(a)* will resemble strain A or *AA(a)*, and *AB(b)* will resemble strain B or *BB(b)*.

Since such effects are solely produced by cytoplasm of the egg, they are described as maternal effects. However, maternal effects are often produced due to effect of nuclear genes through cytoplasm. In other words, properties of cytoplasm depend on nuclear genes. Such cases can be distinguished from those, where extrachromosomal or cytoplasmic hereditary units are present and function either independently or in collaboration with nuclear genetic system. This is called extrachromosomal or cytoplasmic or organellar inheritance, and is distinguished from maternal effects discussed above.

We know that in chromosomes, DNA is the sole genetic material and is the storehouse of genetic information. Discoveries of presence of DNA in cell organelles found outside the nucleus is a strong evidence to suggest that genetic information does exist in cytoplasm also.

It will be seen in this chapter, that two important and essential organelles *i.e.* plastids (in plants only) and mitochondria located in cytoplasm carry DNA. These organelles control extrachromosomal inheritance in many cases through their DNA, which carries genetic information.

However, there are other cases, where cytoplasmic inheritance depends on extra-chromosomal particles, which are not essential for cell function and, therefore, may be present or absent. Such dispensable particles include kappa in *Paramecium,* and sigma in *Drosophila.* These particles are not only inherited but are also infective, since they can be introduced into new hosts without the need of actual process of reproduction.

Further, the presence and reproduction of these infective particles may also depend on nuclear genes. Therefore, in this chapter we will discuss the following three kinds of examples:

(i) Maternal effects depending indirectly on nuclear genes and involving no known cytoplasmic hereditary units,

(ii) Cytoplasmic inheritance involving dispensable and infective hereditary particles in cytoplasm, which may or may not depend on nuclear genes,

(iii) Cytoplasmic inheritance involving essential organelles like chloroplasts and mitochondria.

Maternal Effects

Pigment in Flour Moth (Ephestia kuhniella) : A distinct case of maternal effect was discovered in flour moth *(Ephestia kuhniella)* by Caspari (1936). Dark brown eyes and presence of pigment in other parts of the body in this moth are controlled by a dominant gene *A*, responsible for production of a pigment precursor kynurenine. Homozygous recessive *aa* lacks kynurenine, so that it exhibits absence of pigment and the eyes thereforc have red colour. When heterozygote *Aa* (pigmented) is crossed to nor-pigmented homozygous recessive *aa* (*aa*♀ × Aa ♂) as expected, progeny segregates 1 *Aa* : 1 *aa*, which phenotypically gives the ratio 1 pigmented : 1 non-pigmented. In the reciprocal cross, pigmented *Aa* (♀) × non-pigmented *aa* (♂), the progeny *(lAa : 1 aa)* had all the early larvae pigmented. In this case, however, when larvae matured, only half of them *(Aa)* were dark brown eyed, the other half *(aa)* were red eyed (Fig.). These homozygous *(aa)* pigmented larvae received their egg cytoplasm from mother *(Aa)* and, therefore, had *kynurenine* in early stages of development, but they were incapable of synthesizing their own *kynurenine* in the absence of dominant allele and hence loss of pigment in adult moth.

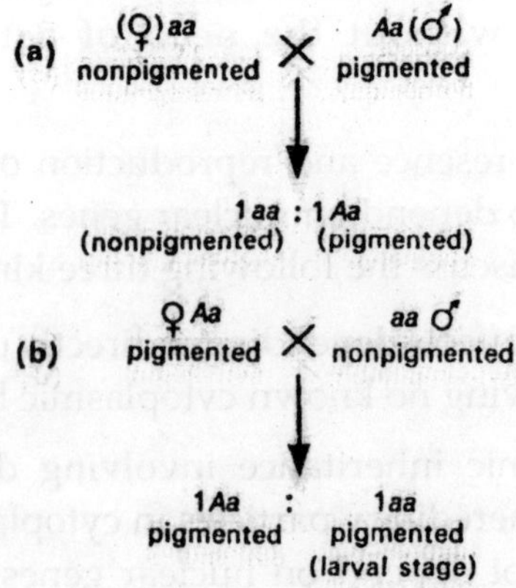

Results of reciprocal crosses (♀ non-pigmented* aa *×* ♂ *pigmented* Aa; ♀ *pigmented* Aa *×* ♂ *non-pigmented* aa*) in flour moth* (Ephestia kuhniella) *showing inheritance of pigment colour.

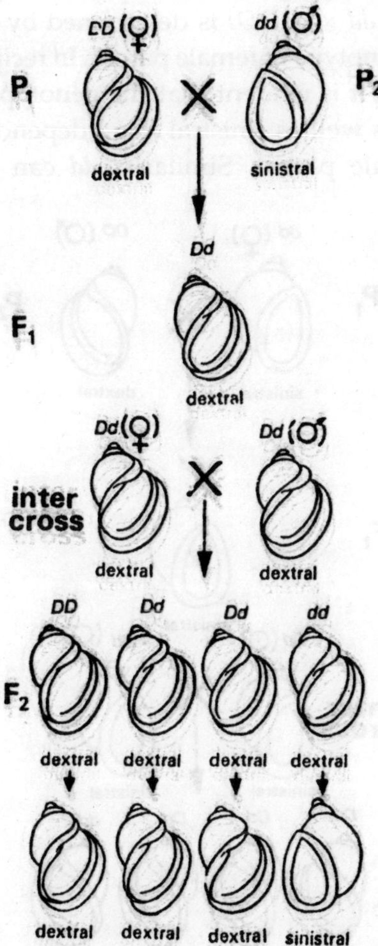

Results in F_1, F_2 and F_3 from a cross between ♀ dextral (DD) and ♂ sinistral (dd) *snails.*

Coiling of Shell in Snail (Limnaea Peregra) : In snails, coiling of shell can be dextral (coiling to the right) or sinistral (coiling to the left). This direction of coiling is genetically controlled, the dextral coiling depending upon dominant allele *D* and sinistral coiling depending upon recessive allele *d*, so that dextral is *DD* and sinistral is *dd*.

The phenotype in progeny obtained from reciprocal crosses (♀ *DD* × ♂ *dd* : ♀ *dd* × ♂ *DD*) is determined by the genotype and not by the phenotype of female parent. In reciprocal crosses shown in Figures, it is evident that the genotype *Dd* (F_1) can be dextral (Fig.) as well as sinistral (Fig.) depending upon the genotype of female parent. Similarly, *dd* can be dextral if

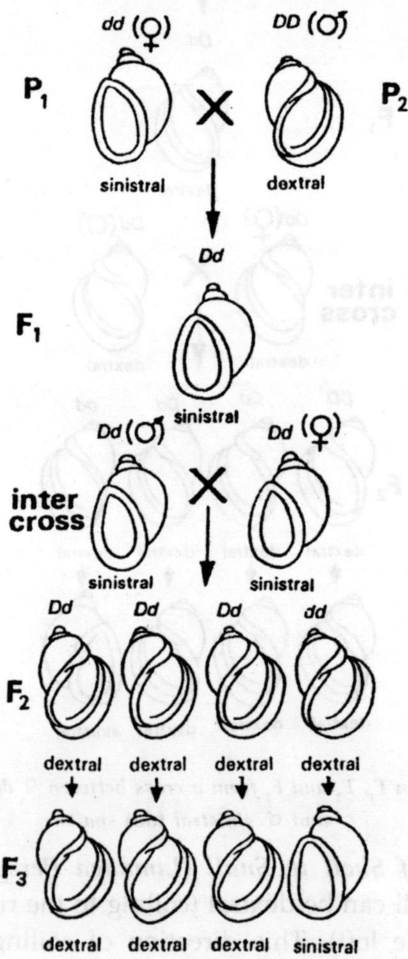

Results in F_1, F_2 and F_3 from a cross between ♀ sinistral (dd) and ♂ dextral (DD) snails.

genotype of female parent carries dominant allele *(Dd)*. It should be carefully noted that phenotype of female parent does not have any effect on phenotype of progeny. It is the genotype of female parent which is really decisive. This is an example of delayed effect of genotype.

Cytoplasmic Inheritance Involving Dispensable Hereditary Units (Nuclear genes may or may not be involved)

Kappa in Paramecium : Kappa is found in certain killer strains of *Paramecium* and are responsible for production of substance paramecin, which is toxic to strains not possessing kappa (sensitive strain). The presence of kappa is dependent

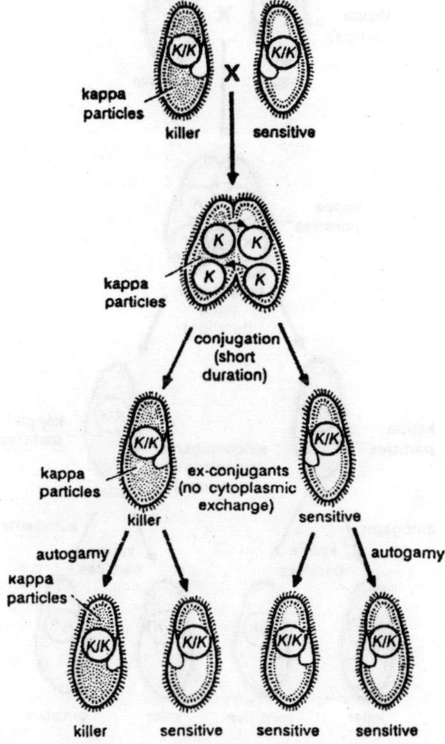

Results of a cross between a killer (KK) and a sensitive (kk) strain of Paramecium, when no cytoplasmic exchange is allowed.

on a dominant allele *K*, so that killer strains are *KK* or *Kk* and sensitive strains are ordinarily *kk*. In absence of dominant allele *K*, kappa cannot multiply and in absence of kappa particles, dominant allele *K* cannot produce them *de novo*. Consequently sensitive strains with genotypes *KK* or *Kk* can be obtained. These will not carry kappa. However, killer strain with genotype *kk* cannot be obtained, because even if kappa is present, it would be lost in absence of dominant allele. If *Paramecium* clones with genotypes *KK* or *Kk* are allowed to multiply asexually at such a fast rate, that multiplication of kappa cannot keep pace with division of cells, kappa will be eventually lost.

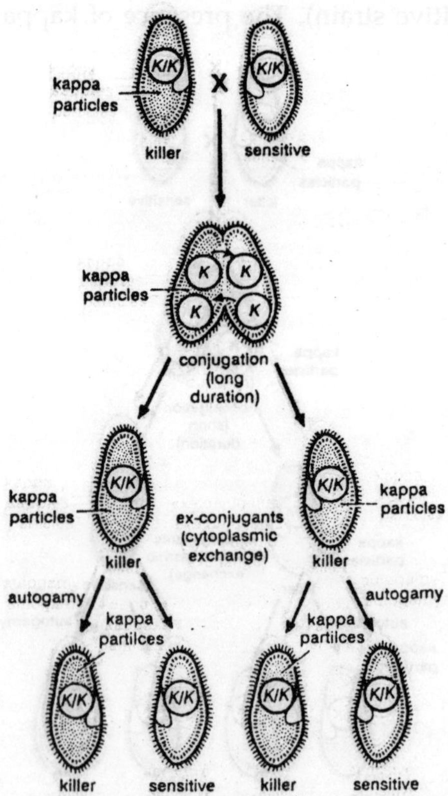

Results of a cross between a killer* (KK) *and a sensitive* (kk) *strain of* Paramecium, *when cytoplasmic exchange is allowed.

Consequently sensitive strains with dominant genotype *(KK, Kk)* having no kappa would be obtained.

If the killer *(KK)* and sensitive *(kk)* strains are allowed to conjugate, all exconjugants (the cells separating after conjugation) will have same genotype *Kk*. Phenotypes of these exconjugants will, however, depend upon duration for which conjugation is allowed. If conjugation does not persist long enough for exchange of cytoplasm, heterozygote *(Kk)* exconjugants will only have parental phenotypes. It means that killers will remain killers and sensitive will remain sensitive even after conjugation (Fig.). If conjugation persists, sensitive strain will receive kappa and will become killer, so that exconjugants will be killers having genotype *Kk* (Fig.).

Killer paramecia was the first example of cytoplasmic inheritance in animals. Although, initially the nature of kappa was not known and it was considered to be a self reproducing gene product, later it was shows to be a symbiotic bacterium. Subsequently, many species of bacterial symbionts were discovered in *Paramecium*, giving examples of free-living organisms entering part of the cell and becoming an integral part of the cell.

CO_2 Sensitivity in Drosophila (Sigma Factor) : Most *Drosophila* flies, can be subjected to a contact with pure CO_2 for long hours, without injury. L'Heritier and Teisser (1958), however, discovered a true breeding strain of *Drosophila* which was sensitive to CO_2. The sensitive flies, when exposed briefly to CO_2 for a short period, become unconscious in a characteristic way, with their legs becoming paralysed. When reciprocal crosses were made between CO_2 sensitive and normal strains, it could be shown that the trait was inherited only from female parent. In other words, while sensitive mothers always give sensitive progeny, sensitive fathers only rarely give sensitive progeny if the mother is normal. In rare transmission through male also, sensitivity is quickly lost after first generation. It has also been shown that if an extract obtained by crushing sensitive flies, is injected into the body of wild (normal) flies, sensitivity can be induced. It was also shewn that this sensitivity can be attributed to a virus like particle called sigma found in cytoplasm of the cells of a sensitive fly.

Sigma factor is transmitted through egg cytoplasm and its reproduction depends on initial supply and on suitable temperature of 20°C, because it is heat labile at high temperature. It does not need a specific gene and may be found associated with different genotypes. A sensitive fly retains its sensitive trait, even when all its chromosomes are replaced by those of normal fly, suggesting that sigma factor or particle is located in cytoplasm and has properties of non-chromosomal genes or plasma genes. Some physical characteristics of *sigma* are also known. It seems to be a particle of 0.07 micron in diameter, and contains DNA responsible for its hereditary nature.

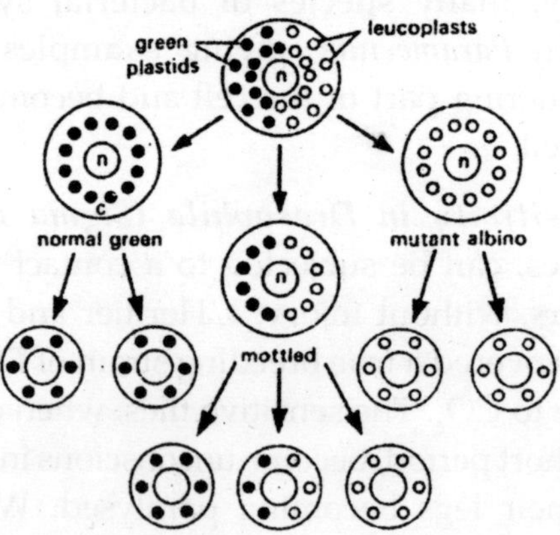

Distribution of green plastids and leucoplasts during gamete formation.

Organellar Genetics

Although most heritable traits in higher organisms are controlled by DNA (deoxyribonucleic acid) in the nucleus, there are now available, a fairly large number of examples where DNA located in cytoplasmic organelles is known to control hereditary traits. DNA was shown to be present in chloroplast (Ris and Plaut, 1962) as well as in mitochondria (Reich and Luck, 1966), and this led to the development of science of organellar genetics. There are a number of well established examples of organellar inheritance available now. Some of the important ones are discussed in this section.

Plastid Inheritance : Variegation in Plants : Most firmly established cases of true cytoplasmic inheritance are those that involve plastids, which are minute cytoplasmic organelles in plant cells. Most important of these plastids are chloroplasts carrying green pigment, the chlorophyll. These chloroplasts duplicate themselves independently of other parts of the cell, and carry some genetic information in the form of DNA. As obvious, usually plastids are transmitted through egg cytoplasm, and only very rarely few, if any plastids are transmitted through pollen. Consequently characters depending on plastids exhibit cytoplasmic inheritance sometimes independent of nuclear genes as in *Mirabilis jalapa*. Plastids also

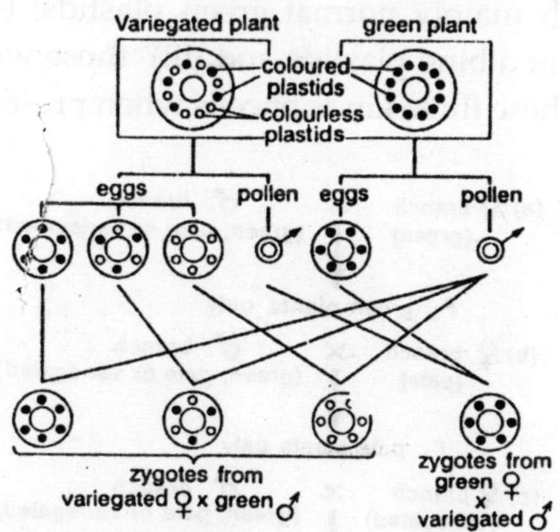

Distribution of coloured and colourless plastids in eggs produced in variegated plant used as ♀ and their effect on progeny obtained due to a cross with normal green plant used as male (♀) parent. The results of a reciprocal cross (green ♀ × variegated (♀) are also shown.

carry DNA which may contain genetic information. Variations due to mutations are also known in these plastids. However, it does not mean that plastid characters are not controlled by nuclear genes, because in a variety of crop plants, albino seeding characters have been found to be under the control of nuclear genes which have also been mapped in some cases. Such cases of nuclear inheritance involving lack of green colour in albino seedlings will not be discussed in this chapter, since these do not concern cytoplasmic inheritance.

It is believed that variation in colour of leaves, branches or whole plants is due to two kinds of plastids (normal and mutant albino). These two types of plastids will faithfully multiply and give rise to their own types due to cell division. But these daughter plastids may not equally distribute themselves to daughter cells. Further, in rapidly dividing cells, division of plastids may not keep pace with cell division, so that distribution of plastids to daughter cells may become a chance phenomenon as illustrated in Figure. A cell having both kinds of plastids may give rise to three types of cells namely (i) those with mainly normal green plastids; (ii) those with mainly mutant albino plastids and (iii) those with both kinds of plastids. These three kinds of cells when present as eggs will

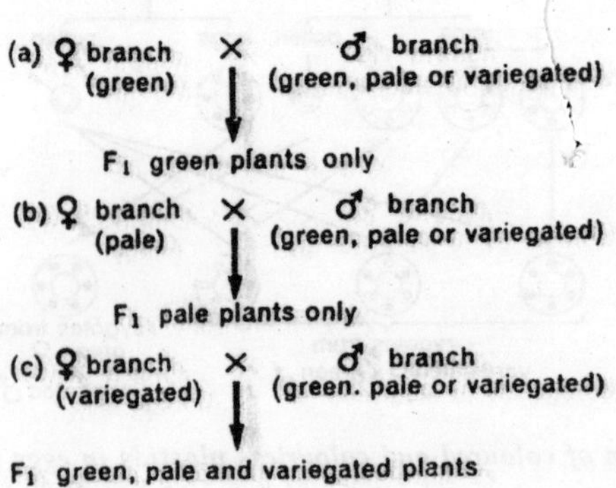

Plastid inheritance in four o'clock showing dependence on the nature of female branch.

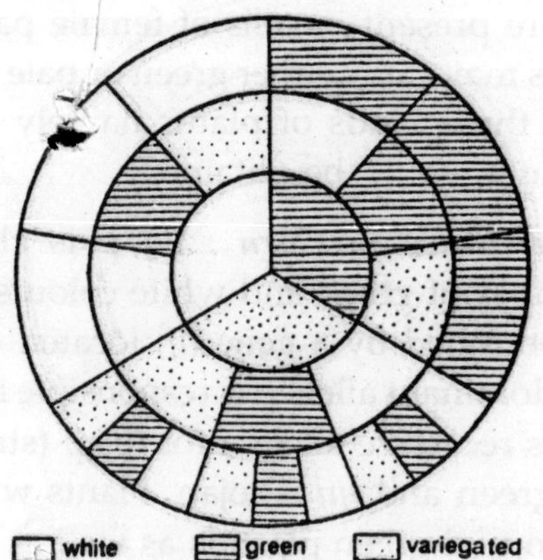

Plastid inheritance in four o'clock. The central circle represents the type of branch that produces flower which are pollinated. Intermediate circle represents branch from which pollen is used and outer circle shows progeny.

give three kinds of progenies. However, these three kinds of cells will not be distinguished as sperms and, therefore, variegated plant when used only as male will give only one kind of progeny (Fig.).

Plastid Inheritance in Four o'clock (Mirabilis Jalapa) : Although a number of cases are now known in plants, where inheritance of plastids is controlled by nuclear genes, there are also cases where this character is transmitted through the agency of cytoplasm alone. Since major part of cytoplasm in zygote is derived from egg, inheritance in such cases will be maternal.

In four o'clock plant (*Mirabilis jalapa*), three kinds of branches with respect to occurrence of plastids may be found. These are (i) completely green, (ii) completely pale green or (iii) variegated. In such cases, phenotype of progeny will depend upon phenotype of branch on which flowers are pollinated (Fig.). Results from three kinds of branches used as female parents are diagrammatically shown in Figure.

As would be evident from Figure and Figure, when variegated branches are used as female source, both green and pale plastids are present in cells of female parent. Therefore, female gametes may carry either green or pale plastids or both. Consequently, three kinds of plants namely green, pale and variegated plants would be obtained.

Iojap Inheritance in Corn : Iojap is characterized by contrasting stripes of green and white colours on leaves. This character is controlled by a gene (*ij*) located on chromosome 7 in corn. The dominant allele *Ij* is responsible for normal green plastids and its recessive allele *ij* for iojap (striped) character, so that *IjIj* is green and *ijij* is iojap. Plants with genotype *ijij* would carry normal green plastids as well as abnormal white plastids, while plants with genotype *Ijij* would carry only green normal plastids.

Female gametes from *ijij* can, therefore, carry either only green plastids or only abnormal white plastids or both. Consequently, if *ijij* is used as female parent and *Ijij* as male parent, the F_1 individuals *(Ijij)* will be green, white or striped. From these F_1 individuals, if *Ijij* (striped) is again used as a female parent, and crossed with *Ijij* (green) as male parent, two genotypes, namely *Ijij* and *Ijij* would be obtained in equal proportions. Each of these two genotypes would have three phenotypes (green, white and striped). It should be noticed that, in such a cross, *Ijij* (striped) would also be obtained (Fig.). It is, therefore, obvious that iojap character once inherited through female parent will be able to maintain itself in the cytoplasm irrespective of genotype of plant. Therefore, as discussed above, iojap or striped phenotype may have any of the three possible genotypes, *Ijij*, *Ijij* and *ijij*.

Male Sterility in Plants

Types of Male Sterility : Male sterility in plants can be controlled by nuclear genes or cytoplasm or by both. Therefore, broadly there are atleast three different mechanisms for control of male sterility in plants. These three types would be briefly discussed in this section:

(a) *Genetic Male Sterility* : In this type, male sterility is controlled by a single gene and is recessive to fertility, so that the F_1 individuals would be fertile. In the F_2 generation, the fertile and sterile individuals will segregate in 3 : 1 ratio (Fig.).

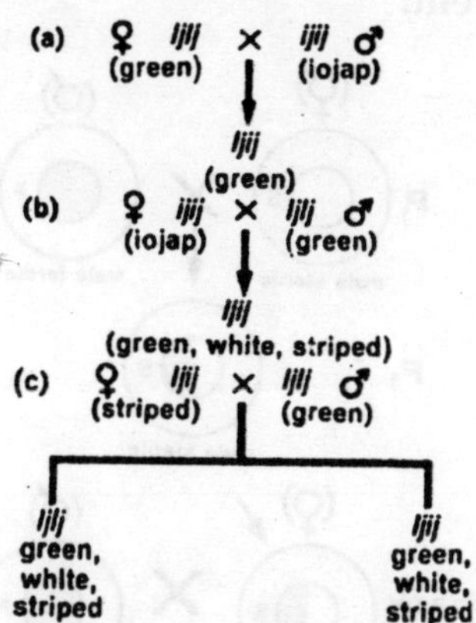

Inheritance of Iojap striping in maize showing dependence on the phenotype of female parent.

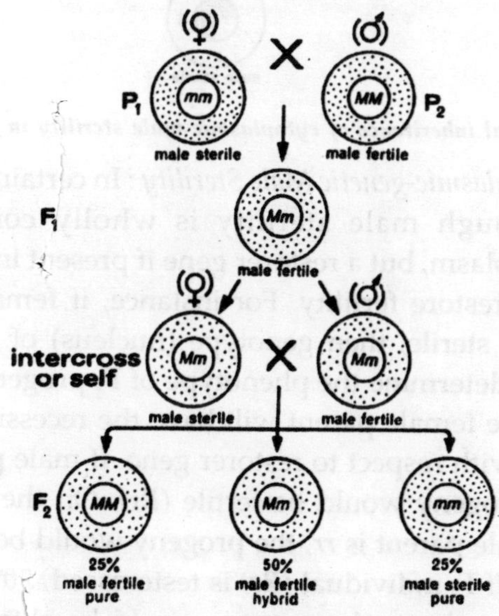

Inheritance of genetic male sterility in plants.

(b) *Cytoplasmic Male Sterility* : In several crops like maize, cytoplasmic control of male sterility is known. In such cases if female parent is male sterile, F_1 progeny would always be male sterile (Fig.), because cytoplasm is mainly derived from egg obtained from male sterile female parent.

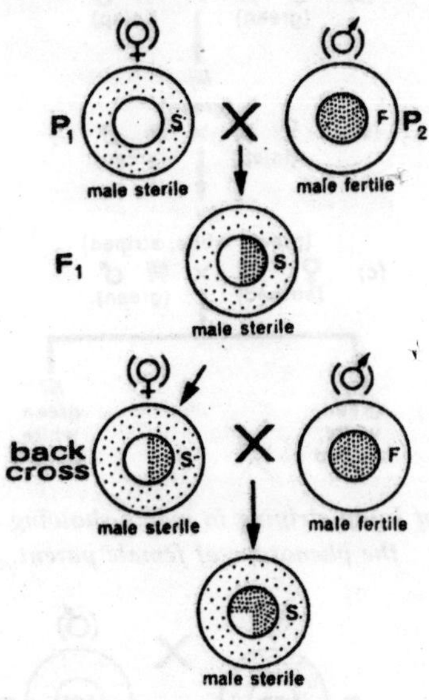

Maternal inheritance of cytoplasmic male sterility in plants.

(c) *Cytoplasmic-genetic Male Sterility* : In certain other cases, although male sterility is wholly controlled by cytoplasm, but a restorer gene if present in the nucleus will restore fertility. For instance, if female parent is male sterile, then genotype (nucleus) of male parent will determine the phenotype of F_1 progeny. The male sterile female parent will have the recessive genotype *(rr)* with respect to restorer gene. If male parent is *RR*, F_1 progeny would be fertile *(Rr)*. On the other hand, if male parent is *rr*, the progeny would be male sterile *(rr)*. If F, individual *(Rr)* is testcrossed, 50% fertile and 50% male sterile progeny would be obtained (Fig.).

During 1990-92, transgenic male sterile and fertility restorer plants have also been produced in *Brassica napus*, by transfer of *'barnase'* and *'barstar'* genes from *Bacillus amyloliquefaciens*.

Cytoplasmic Male Sterility in Maize : Rhoades in 1933, reported the analysis of first cytoplasmic male sterile plants in maize and demonstrated that male sterility was contributed by female parent and that nuclear genes had no influence. This

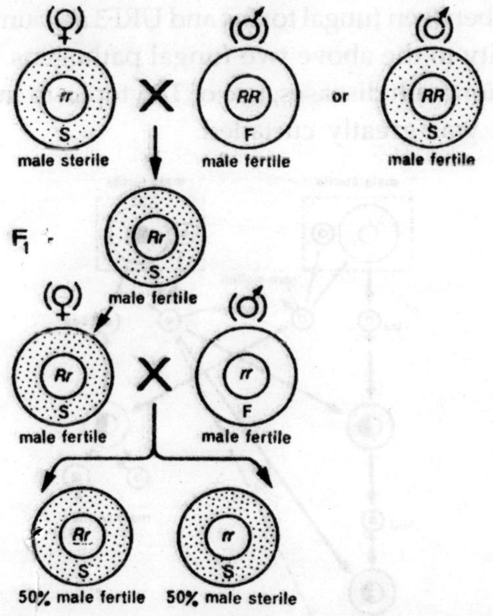

Inheritance of cytoplasmic male sterility and restoration of fertility due to restorer gene.

was shown by crossing male sterile plants with wide range of fertile males and by observing that in subsequent generations all progenies were male sterile.

In maize, three male sterile sources (cms) are known, which are called T, C and S. The normal male fertile cytoplasm is called N cytoplasm. Each of the three cms cytoplasms exhibits strict maternal inheritance and even when all chromosomes were replaced by a male fertile source through backcrosses, male sterility could not be overcome (Fig.). These cytoplasms were a great asset in production of hybrid corn seed at commercial scale. It was also shown that T (Texas) cytoplasm was associated with (i) susceptibility to two diseases, namely southern corn leaf blight disease *(Bipolaris maydis* race T, formerly known as *Helminthosporium maydis)* and yellow leaf blight *(Phyllstica maydis)*; (ii) an unusual mitochondrial gene *T-rifle*, which encodes a 13 kilodalton polypeptide (URF13). An interaction between fungal toxins and URF3 accounts for specific susceptibility to the above two fungal pathogens. Due to such an association with diseases, use of T-cytoplasm in hybrid seed production was greatly curtailed.

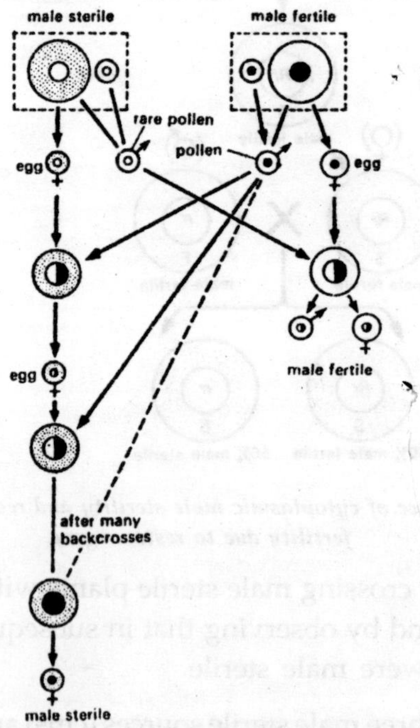

Reciprocal crosses between male sterile and male fertile plants (using rare pollen from male sterile). Figure shows a number of back crosses of male sterile as female parent, substituting the nucleus from male fertile. Note that the substitution of male fertile nucleus does not restore fertility in male sterile.

(a) *Role of Mitochondria :* Recent studies have proved that factors responsible for cytoplasmic male sterility are located in mitochondrial DNA (mtDNA). Six kinds of evidences proved mitochondrial role, (i) When mitochondrial DNA was studied in three male sterile cytoplasms (T, C and S) and normal male fertile (N) cytoplasm, it was found that mtDNAs differed in four cytoplasms. It is also apparent that mtDNA from a particular cms (cytoplasmic male sterile) source is stable and conserved. For instance Texas (T) cytoplasm, even after 20 generations of backcrosses had same mtDNA from different T sources. However, when chloroplast DNA (ctDNA) was similarly studied, very little, if any differences were found between cpDNA from four different sources (N, T, C, S). Such an observation was used as an evidence to suggest that cytoplasmic male sterility in higher plants may be associated with mtDNA. (ii) Association of mtDNA has also been shown with disease susceptibility, which itself is associated with cytoplasmic male sterility (cms). Both pathogens, one causing leaf blight disease and the other causing yellow leaf blight are known to produce host specific *toxins.* These toxins affect the mitochondria of T cytoplasm, but not of N cytoplasm. Such an effect of toxins on mitochondria was observed both inside the cell *(in vivo* systems) and in cell free mitochondrial preparations *(in vitro* systems), (iii) The effect of toxins on mitochondria was modified in cytoplasmic male steriles, which were restored to fertility with the help of restorer genes *(*Rf_1, Rf_2-nuclear genes; if these genes are present, male sterile cytoplasm does not express itself)- This suggests that restorer genes help restore fertility through their action on mitochondria, (iv) When T-cytoplasm cells in culture were treated with toxin and resistant cells were selected, then mitochondria of these resistant cells were unaffected by toxin. Mature plants derived from these resistant cells selected originally from male sterile plants, were fertile a disease resistant. It could be shown that this ch involved changes in mitochondria only, (v) It ha shown in several organisms that there are proteins synthesized under the control of mtDI these proteins include cytochrome oxidase, cytoc b and ATPase. There are evidences available tha cytoplasm, *cytochrome oxidase, cytochrome b* and get affected. Cytoplasm specific proteins character of individual cytoplasms have been shown to be pres in N, T, C and S cytoplasms, although the sites of acti or function of these distinguishing proteins are nc known, (vi) When microsporogenesis was studied in anthers of T and N cytoplasms, mitochondrial degeneration was observed in tapetum and middle layers of anther at the tetrad stage in T cytoplasm and not in N cytoplasm. This degeneration includes, swelling of mitochondria, loss of cristae and internal disorganization. Contrary to this, plastids and other organelles were not affected.

(b) *Role of Plasmid Like DNA in S-cytoplasm* : Cytoplasmic male sterility due to S-cytoplasm has been shown to be different from T cytoplasm in several ways. The mtDNA preparations from S-cytoplasm were found to contain two unique plasmid (plasmids are DNA fragments resembling viral DNA molecule) like DNAs called SI (6397 bp) and S_2 (5453 bp). These plasmid like DNAs could not be isolated from DNA preparations from chloroplasts or nuclei, showing that these are characteristic of mitochondria. Further, the plasmid like DNAs are not found in N (fertile), T (sterile) and C (sterile) cytoplasms, but were present in S-cytoplasm with a variety of nuclear backgrounds, suggesting that these plasmid like DNAs may be responsible for causing male sterility associated with S-cytoplasm.

T and C sterile cytoplasms have been found to be stable, never giving rise to fertile cytoplasms even by applying utagens. However, S-cytoplasm was found to give rise to condition in some cases due to one of the two kinds of ;es : (i) cytoplasmic mutation from male sterile to male e or (ii) nuclear mutation giving rise to a new restorer . Both these changes could be involved, as shown from periments when these fertilise derived from S-cytoplasm were crossed as males with cms-S testers. In some cases the progeny was male sterile, suggesting the absence of restorer gene, while in other cases the progeny was semi-fertile suggesting the presence of restorer gene. In the latter case, restorer was found to be different from the normal nuclear restorer gene Rf_3 meant for S-cytoplasm. These new restorers could be located on different chromosomes, which led to the belief that a fertility element having characteristics of an *episome* may be present, which gets attached to different chromosomes at different times leading to change from sterile to fertile condition. These fertility elements may actually be the plasmid like DNAs.

It has also been suggested that male fertility genes could be originally located on organellar DNA and were later transposed to a nuclear site giving rise to restorer gene. When this fertility gene is absent both from nucleus and organelle, this might have led to cytoplasmic male sterility. Such a hypothesis for origin of cytoplasmic male sterility is being examined now.

Chloroplast Genetics

Non-chromosomal Genes in Chlamydomonas : In *Chlamydomonas*, although the first non-Mendelian gene controlling streptomycin resistance was described by Ruth Sager (a lady scientist in U.S.A., who died in 1997) as early as

1954, but its location on chloroplast genome was known later. Several gene pairs located on chloroplast genome were reported and a circular genetic map was proposed by Sager in 1972 in her book *"Cytoplasmic Genes* and *Organelles"*. Later she presented further results and discussed methods used in the geneti analysis of these non-Mendelian genes.

(a) *Uniparental Inheritance* : As shown earlier, se characters show maternal inheritance both in p and animals. This was earlier explained to be du greater amount of cytoplasm in egg than in sperm male gamete. In *Chlamydomonas,* which is a haploid organism, sexual reproduction is isogamous, both gametes being similar in morphology and thus having equal amounts of cytoplasm. Despite this lack of difference in cytoplasmic content, in *Chlamydomonas,* some characters are transferred only from one parent, exhibiting what is called uniparental inheritance, rather than maternal inheritance, because in *Chlamydomonas,* we describe two sexes as two mating types, *i.e., mt^+* (plus) and *mt^-* (minus), rather than as female and male (Fig.). One may notice that in diploids, the male gametes do not contribute cytoplasm (Fig.). One may also

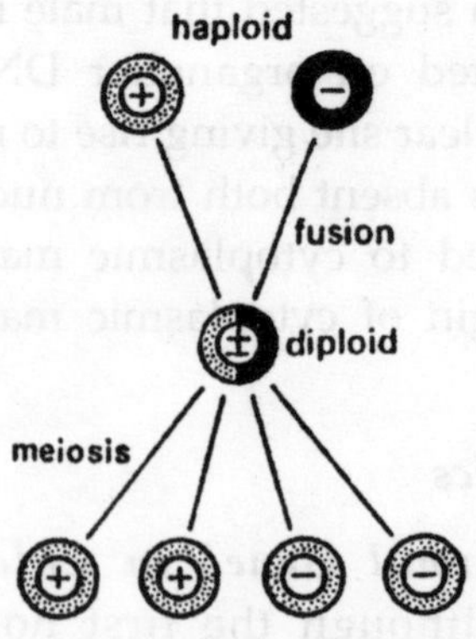

Uniparental inheritance in a haploid organism like* Chlamydomonas, *although cytoplasm is transmitted from both the gametes.

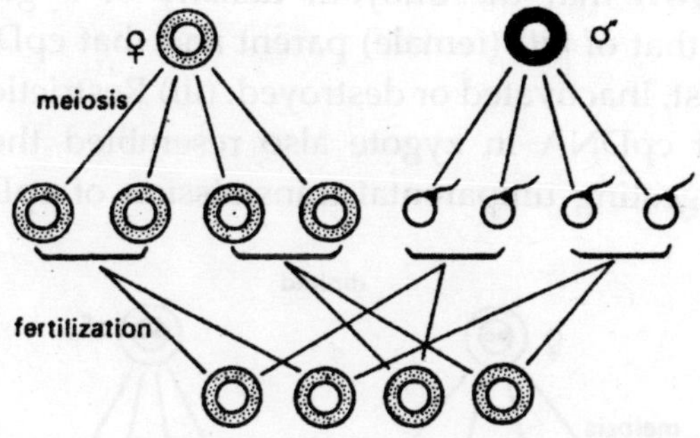

Maternal inheritance in a diploid organism, where cytoplasm is transmitted only through the egg.

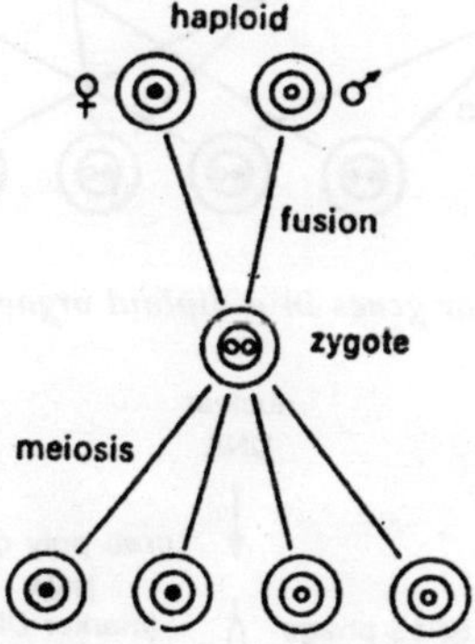

Segregation of nuclear genes in a haploid organism (1:1 ratio).

distinguish between the patterns of inheritance between non-chromosomal and chromosomal genes. The inheritance patterns for chromosomal genes in haploid and diploid organisms are shown in Figures.

Uniparental inheritance has also been demonstrated at the level of cpDNA, due to the following facilities : (i) cpDNA forms a separate distinct band on CsCl density gradient centrifugation (Fig.); (ii) the cpDNA of two parents can be differently labelled with N^{14} and N^{15}, the two giving buoyant densities of 1.69 and 1.70 respectively; this difference looks small, but they form separate bands on CsCl density gradient.

It was shown that the buoyant density of zygote always resembled that of mt^+ (female) parent and that cpDNA of mt^- parent is lost, inactivated or destroyed, (iii) Restriction enzyme patterns of cpDNA in zygote also resembled those of mt^+ parent suggesting uniparental transmission of cpDNA.

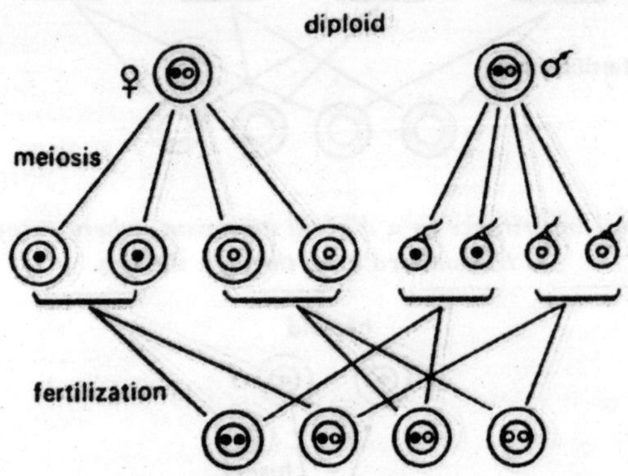

Segregation of nuclear genes in a diploid organism (1 : 2 : 1 ratio)

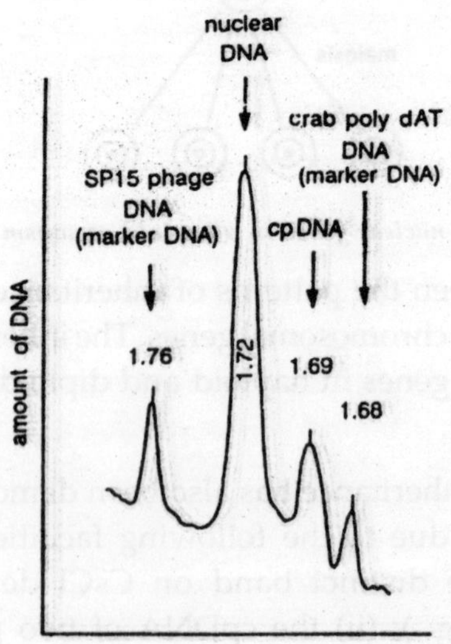

CsCl density gradient centrifugation of Chlamydomonas *DNA, showing cpDNA band distinct from nuclear DNA band.*

(b) *Segregation, Independent Assortment and Gene Mapping* : In earlier years (before 1960), a large number of cases were known, where non-chromosomal inheritance, as evident from uniparental or maternal inheritance, were described. However, further detailed study of these traits was not possible. There were two reasons for this. *First,* the mutant non-chromosomal genes could not be artificially produced when desired and spontaneous mutations for these genes were rare, thus making their study difficult. *Second,* since these traits were transmitted from only one parent, standard techniques of genetic analysis could not be applied to study segregation and independent assortment as studied in chromosomal genes. Ruth Sager was successful in overcoming both these difficulties in *Chlamydomonas* and was thus able to study the non-chromosomal genes in this organism in some detail.

Ruth Sager found that if streptomycin sensitive cells are grown in a medium containing streptomycin, several kinds of induced mutations in Chloroplast including streptomycin resistant mutations were produced. Streptomycin resistance could be found on chromosomes also where it may be inherited in a Mendelian manner, but in some strains this character showed uniparental inheritance as shown in Figure. It may be noticed that when the mt^+ strain was streptomycin resistant, all the progeny was streptomycin resistant and when it was sensitive, all the progeny sensitive. Streptomycin was used for inducing not only resistance to streptomycin but also to induce mutations for several other non-chromosomal traits thus making it possible to study their inheritance.

The second difficulty of getting the transmission of non-chromosomal genes from only one parent was overcome when it was observed by Ruth Sager that in very rare cases streptomycin resistance and streptomycin sensitivity from parents of both mating types, mt^+ and mt^- could be transmitted.

It was found that, when streptomycin dependent strain *(sd,* which will not survive without streptomycin) was used as *mt*⁺ strain, and progeny grown in a medium containing streptomycin, only *sd* strains will be produced (Fig.). However, if the same progeny is grown in a medium lacking streptomycin, then biparental exceptional zygotes will grow, because the uniparental *sd* strains will not survive in the absence of streptomycin (Fig.). Similarly in the reciprocal cross, using *ss* (streptomycin sensitive) as the mating type *mt*⁺ and *sd* as *mt*~, only *ss* strains will be available in the absence of streptomycin showing uniparental inheritance (Fig.), but, if the progeny is grown in the presence of streptomycin, then biparental zygotes will be produced (Fig.). Later it was discovered that if *mt** parent is irradiated with ultraviolet (UV) irradiation prior to mating, frequency of biparental

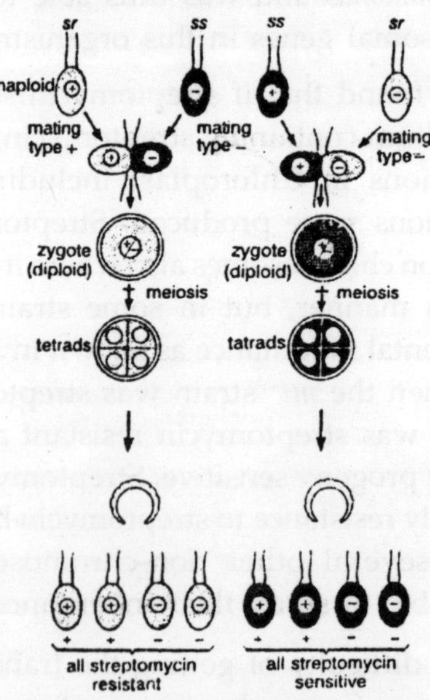

Reciprocal cross* (sr mt⁺ × ss mt⁻ *and ss mt⁻ × sr mt⁻) showing uniparental inheritance and lack of segregation of non-chromosomal genes (sr = streptomycin resistance and ss = sensitivity to streptomycin) along with segregation of nuclear genes (mt⁺ and mt⁻).

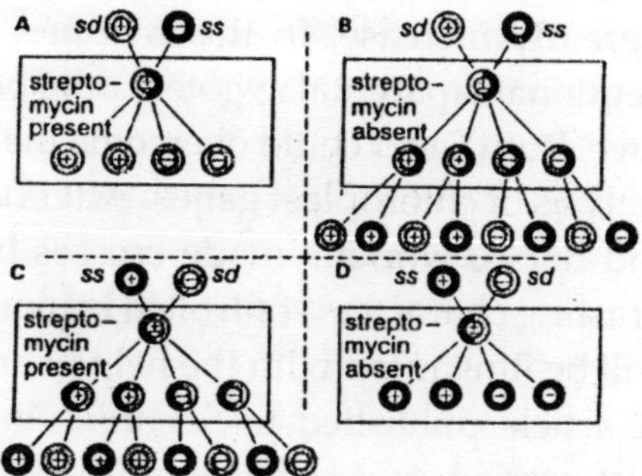

Results of reciprocal crosses (sd mt⁺ × ss mt⁻; ss mt⁻ × sd mt⁻), when the progeny is grown under two conditions (streptomycin present or streptomycin absent). Unfavourable conditions (absence of streptomycin in B and presence of streptomycin in C) lead to selection of exceptional biparental zygotes.

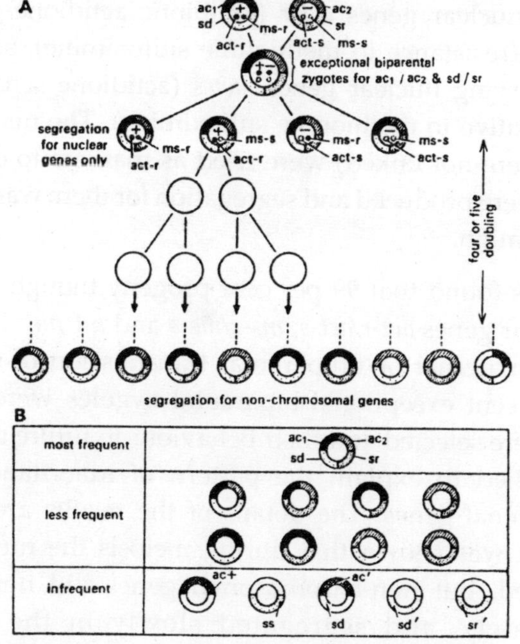

A cross involving two non-chromosomal (ac_1/ac_2 and sd/sr) and three nuclear genes (act-r/act-s; ms-r/ms-s; mt^+/mt^-) showing segregation of nuclear genes soon after meiosis and that of non-chromosomal genes in the following generations (A); relative frequencies of different cytoplasmic types obtained in this cross (B); ac^+, ac′, ss, sd′, sr′ represent new types (result of only 1% exceptional biparental zygotes..

zygotes will greatly increase. In this manner, techniques for obtaining exceptional biparental zygotes could be standardized. As shown above, Ruth Sager could overcome the two difficulties for genetic analysis of chloroplast genes. After having achieved this, Sager and her co-workers made crosses between strains carrying contrasting characters controlled by non-chromosomal genes. This will be illustrated with the help of an example used by her in her article published in *Scientific American* in 1965. The mt^+ parent carried ac_1 (acetate requiring mutant) and *sd* (streptomycin dependence), and *mt* parent carried ac_2 and *sr* (the other alleles of same genes ac_1 and *sd).*

The parents also differed in two other nuclear genes, mt^+ carrying nuclear genes *act-r* (antibiotic actidione resistance) and *ms-r* (resistance to methionine sulfoximine), and the mt^- strain carrying nuclear genes, *act-s* (actidione sensitive) and *ms-s* (sensitive to methionine sulfoximine). The nuclear genes (which were not linked) were used as markers to ensure that zygotes were produced and segregation for them was according to expectation.

It was found that 99 per cent progeny though segregated for nuclear genes *act-r/act-s, ms-r/ms-s* and mt^+/mt^+ but they all showed maternal or uniparental transmission of ac_1 and *sd*. One per cent exceptional biparental zygotes were obtained, which were selected and their behaviour in future generations was studied to explain the pattern of inheritance of non-chromosomal genes. The details of the results are shown in Figure. It was shown that during meiosis the nuclear genes segregated, but non-chromosomal genes did not segregate immediately, and segregated slowly in the following generations. The following conclusions were made on the basis of the results shown in Figure. (i) The non-chromosomal or chloroplast genes segregate like nuclear genes in 1 : 1 ratio in mitotic divisions, so that ac_1 and ac_2 as well as *sd* and *sr* behaved as alleles. (ii) The ac/ac_2 and *sd/sr* gene pairs segregate independently of each other, in both time and space, giving rise to parental combinations *ac*$_1$ *sd* and *ac*$_2$ *sr* as well as the recombinants *ac*$_1$ *sr* and *ac*$_2$ *sd*. (iii) Even wild type or new cytoplasmic alleles arose, which are designated as *ac*$^+$, *ac′*, *ss*, *sd′* and *sr′*. These new alleles are believed to have originated due to intragenic recombination.

Several other non-chromosomal genes were studied using the above and other techniques. Some of these genes include the following : (i) *ac* (acetate requiring mutant) with reduced ability to grow and giving small flat colonies; (ii) *Sm*^{2}, streptomycin resistance; there is another gene symbolised as *sd*, which confers streptomycin dependence, so that the organism is not only resistant, but is dependent and can not grow without the drug; (iii) *ery*, erythromycin resistance; (iv) *spc*, spectinomycin resistance; (v) *tntj*, temperature sensitive, which grows only at 25°C and not at 35°C.

A circular genetic map of chloroplast genome of *Chlamydomonas* is shown in Figure, where an attachment point *ap* is also shown. (For details consult Ruth Sager's book).

Chloroplast genetics in higher plants. We know that chloroplasts of lower eukaryotes and higher plants like *Zea mays* contain DNA. Chloroplast genomes are relatively large (with a narrow range) usually ranging from 120 kb to 210 kb (kb = kilo base pairs) except in *Acetabularia* (400 kb). This is

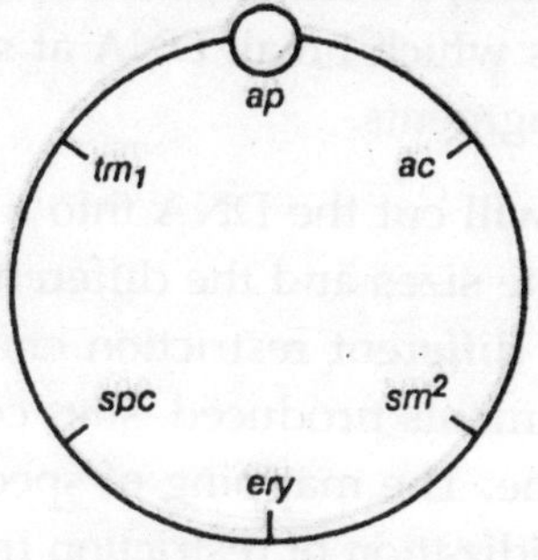

A circular genetic map of Chloroplast genome in **Chlamydomonas.**

comparable to the size of a large bacteriophage like T_4 (110 kb). Each cell may have several chloroplasts and each chloroplast may have several circular DNA molecules, so that the chloroplast DNA (cpDNA) may make up to 14% of cellular DNA in lower eukaryotes.

In higher eukaryotes, however, the cpDNA makes a small proportion of total DNA in the cell. These features of Chloroplast DNA are presented in Table.

Features of Chloroplast DNA (cpONA) in Lower Eukaryotes and Higher Plants

Species	*cpDNA (in kb pairs)*	*No. genomes per organelle*	*No. organelles per cell*	*cpDNA (as % total DNA)*
C. reinhardii	195	70-100	1	11%
Euglena gracilis	135	40	15	3%
Higher plants	120-210	variable	variable	variable

There is evidence that genes controlling specific traits or those controlling the synthesis of specific molecules of RNA and protein, are found on the chloroplast DNA. The presence of these genes could be shown by a technique called restriction fragment analysis or restriction site mapping. In this technique, using specific enzymes, the endonucleases, also called restriction enzymes (enzymes which break DNA at specific sites), DNA can be cut into fragments.

Each enzyme will cut the DNA into a definite number of segments of definite sizes and the different enzymes differ in their effect. Using different restriction enzymes and through an analysis of fragments produced, sites could be mapped on Chloroplast genome. The mapping of specific genes was also facilitated by hybridization of restriction fragments with RNA transcripts, produced in the cell and suspected to have

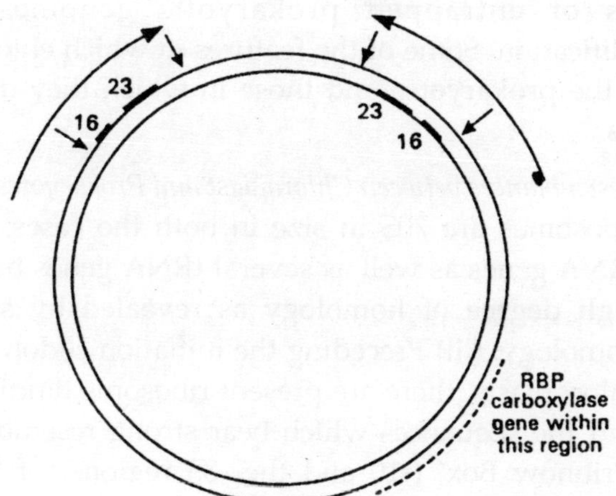

Map of circular chloroplast DNA of* Zea mays, *showing the positions of duplicate inverted region of rRNA genes and also RBP (ribulose biphosphate) carboxylase gene (after Bedbrook et al., 1979).

originated from chloroplast genome. In this manner, small and large rRNA (ribosomal RNA) could be shown to be synthesized by specific genes on chloroplast DNA (cpDNA). In maize, the sequence of rRNA gene is shown to be duplicated. Similarly genes for tRNA (transfer RNA) have also been mapped.

The genes on cpDNA synthesize not only rRNA and tRNA, but also some proteins through the synthesis of mRNA (messenger RNA). Ribulose bisphosphate carboxylase (rubisco) is one such important enzyme, which is abundant in green leaves. Each molecule of this enzyme has 16 polypeptides, eight small (100 amino acids long) and eight large (450 amino acids long). Although small subunit is synthesized through nuclear genes, the bigger is under the control of cpDNA. The map of maize chloroplast genome known at present is shown in Figure.

Chloroplast Genome and Prokaryotes (A Comparison) : A detailed study of chloroplast genome in eukaryotes, has revealed that the chloroplast genomes share both prokaryotic and eukaryotic features. Therefore, the view that they are the remnants of entrapped prokaryotic genomes is an oversimplification. Some of the features in which chloroplasts resemble the prokaryotes and those in which they differ are as follows..

(a) *Resemblances between Chloroplast and Prokaryotes :* (i) The ribosomes are 70S in size in both the cases, (ii) The rRNA genes as well as several tRNA genes have very high degree of homology as revealed by sequence homology. (iii) Preceding the initiation codon AUG in chloroplasts, there are present ribosome binding sites, and the sequences which bear strong resemblance to 'Pribnow Box' (-10 and the -35 regions) of bacteria. (iv) There is sequence overlapping between some genes in chloroplasts, a feature known in prokaryotes. In chloroplasts it can not be due to economy since non coding sequences are also available in chloroplasts. (v) There is a resemblance between RNA polymerases of chloroplasts and prokaryotes. For instance, in *Chlamydomonas reinhardii,* the enzyme is sensitive to rifampicin and contains a (sigma) like polypeptide. There is also evidence of homology between P and P' subunits of *E. coli* RNA polymerase and those of C. *reinhardii* chloroplast RNA polymerase. (vi) Protein synthesis in chloroplasts can be inhibited by chloramphenicol as in bacteria, since only 70S ribosomes are sensitive to this drug. Cycloheximide does not inhibit protein synthesis in chloroplasts, as it does in the cytoplasm (80S ribosomes) of eukaryotes. (vii) Initiation of protein synthesis takes place by formyl-methionyl tRNA both in chloroplasts and prokaryotes.

(b) *Differences between Chloroplast and Prokaryotic Genomes:* (i) In chloroplasts there are split genes, *e.g.* rRNA genes and protein genes in chloroplasts of C. *reinhardii*, protein genes in chloroplasts of *Euglena* and tRNA genes of several chloroplasts of higher organisms. The introns of these split genes fall in three groups and differ from those in nuclear genes of eukaryotes. (ii) Chloroplast DNA sequences have the ability to promote autonomous replication. This feature is absent in prokaryotes. (iii) Presence of short repetitive elements interspersed throughout several, but not all chloroplast genomes is also a eukaryotic feature.

Mitochondrial Genetics : We know that like the chloroplasts, the mitochondria of both animals and plants also contain DNA. But unlike chloroplast genome mitochondrial genome (mtDNA) shows fairly large range of variation in size and in its proportion in the cellular DNA. It varies in size from 15 to l9kb in diptera, mammals, amphibians and fishes; from 15 to l76kb in protists and fungi; from 200 to 2400-kb in higher plants. However, the number of mitochondria per cell is large, although relative to the nuclear DNA, usually the mtDNA still represents only about less than 1% of cellular DNA, except in yeast *(Saccharomyces cerevisiae)*. In yeast, mtDNA may represent as high as 18% of the cellular DNA, each mitochondrial genome being 85kb. Some details of mtDNA are shown in Table.

Mitochondrial DNA in Some Eukaryotes

Species	*mt DNA (kb pairs)*	*No. genomes per organella*	*No. organdies per cell*	*mt DNA (% Total DNA)*
Mouse (L Cell)	16.2	2	500	0.2%
Human (HeLa)	16.8	2.6	750	0.5%
Fruit fly	18.4	Unknown	Unknown	Unknown
Xenopus laevi	18.4	Unknown	Unknown	Unknown
S. cerevisiae	75.0	4	22	18%
Brassica campestris	218.0	1	Unknown	Unknown
Spiruicea oleracea	327.0	1	Unknown	Unknown
Zea mays	570.0	6	Unknown	Unknown
Cucumis melo	2500.0	Unknown	Unknown	Unknown

Mitochondrial DNA, like chloroplast DNA is usually circular, but in cilitate protozoans, *Chlamydomonas reinhardii*, and also in some yeasts, the mitochondrial DNA is a linear molecule. The genetic role of this DNA (mtDNA) is also understood now. The most important work on mitochondria was done in yeast which was initiated by the discovery of petite mutants by B. Ephrussi. Subsequently mtDNA was studied in several organisms including plants and human beings.

'Petite' in Yeast, a Mitochondrial Character : The petite mutants in yeast fail to grow on carbon source like glucose and produce smaller colonies when grown on sugars like glucose. This is a difference which can be observed only in the presence of oxygen, so that these petites actually have a defective aerobic respiratory mechanism. These petites differ from wild type, called grande and are characterized by (i) their insensitivity to inhibitors of aerobic pathways (like cyanide), (ii) absence of cytochromes a, a_3, b and a number of other changes in mitochondrial respiratory enzymes, (iii) incompletely developed state of mitochondria and (iv) lack of stainability of petite mitochondria.

The petite mutants can be 'segregational' following Mendelian segregation and, therefore, presumably controlled by chromosomal genes or may be more often 'vegetative' (non-segregational) or extra-chromosomal. The genetic basis of petite character is a cytoplasmic factor ρ^+ (rho), which may be absent or defective in petites. The vegetative petites, thus can be neutral (ρ^0), which completely lack ρ^+ or they can be suppressive (ρ^-) having a defective ρ^+. The neutral petites are not transmitted while suppressive petites are transmitted to a fraction of vegetative diploid progeny. Suppressiveness varies from 1% to 99% petites in various strains.

Two lines of evidences suggest association of ρ^+ with mitochondrial DNA (mtDNA). (i) Ethidium bromide, which induces petite mutations with 100% efficiency, causes degradation of mtDNA after prolonged exposure of cells. Neutral petites have actually been found to lack mtDNA. (ii) Suppressive petites contain mtDNA, that is often grossly altered in base composition with respect to wild mtDNA.

Poky Neurospora, A Mitochondrial Character : A poky mutant in *Neurospora* differs from wild type in the following attributes : (i) it is slow growing, (ii) it shows maternal inheritance and (iii) it has abnormal cytochromes (cytochromes are proteins needed for oxidation of food and generation of ATP energy). Of the three cytochromes, cyt a, b and c, all the three are found in wild type, while cyt a and b are absent and cyt c is in excess in poky mutant. In reciprocal crosses, poky character shows maternal inheritance (poky ♀ × wild type ♂ → all poky; wild type ♀ × poky ♂ → all wild type), although other marker nuclear genes (e.g. cut^+/ad^-) show 1 : 1 segregation. Following evidences suggested that poky may be located in mitochondrial genome : (i) slow growth may be due to lack of ATP energy and source of this energy is mitochondria; (ii) cytochromes in poky differ from those in wild type in quality and quantity and these cytochromes are found in mitochondria.

Mitochondrial Genetic Markers in Yeast (Saccharomyces Cervisiae)

Phenotype	*Symbol of gene*	*Function*
A. Drug Resistance		
1. Oligomycin	oli^R-*1*	Mt-ATPase, subunit 9
	oli^R-*1*	Mt-ATPase, subunit 6
2. Antimycin and miakmycin	ana^R	Mt-membrane component (cytochrome b ?)
3. Erythromycin	ery^R	Ribosome structure (21S rRNA)
4. Chloramphenicol	cap^R	Ribosome structure (21S rRNA)
5. Paromomycin	par^R	Ribosome structure (15S rRNA)
B. Respiratory, Deficiency		
	oxi-1	Cytochrome oxidase peptide 3

Contd...

Phenotype	Symbol of gene	Function
	oxi-2	Cytochrome oxidase peptide 2
	oxi-1	Cytochrome oxidase peptide 1
	cob	Cytochrome b/coenzyme Q-cytochrome c-reductase
	pho-1, 2	Deficiency in oxidative phosphorylation

Mapping of Mitochondrial Genes in Yeast : The mitochondrial genes in yeast (Table) have been mapped using a variety of techniques, some of them similar to those used for chloroplast genes. These techniques have been extended for mapping mitochondrial genes in other organisms also and will be briefly described. (a) *Recombination Mapping :* Since there are no chloroplasts in yeast, all genes showing uniparental inheritance can be attributed to mitochondria. Uniparental inheritance can be established, when two mating types *(Mat* α and *Mat a)* are fused, the fusion product being a diploid (2n) and also a cytohet (cytoplasmically heterozygous). If diploid cells are allowed to divide mitotically (budding), segregation and recombination occur, but when this budding is followed by meiosis (meiosis can be induced by shifting cells into a special medium), all ascospores in an ascus are one type with respect to mitochondrial genes, although they show segregation (1 : 1) for nuclear genes including the gene for mating type. This is illustrated in Figure using a cross *eryR* × *erys* (erythromycin resistant x erythromycin susceptible).

Using the above principle, if a dihybrid cross such as ery^R spi^R × ery^S spi^S (*spi* = spiromycin) is made and diploid cells are allowed budding followed by sporulation (meiosis), all asci showed lack of segregation (Fig.) and the results like the following could be obtained : ery^R spi^R = 63 tetrads; ery^S spi^S = 48 tetrads; ery^R spi^R= 7 tetrads; ery^R spi^S = 1 tetrad. Among these four, two classes are recombinants, but they result from several rounds of recombination during budding as in phages.

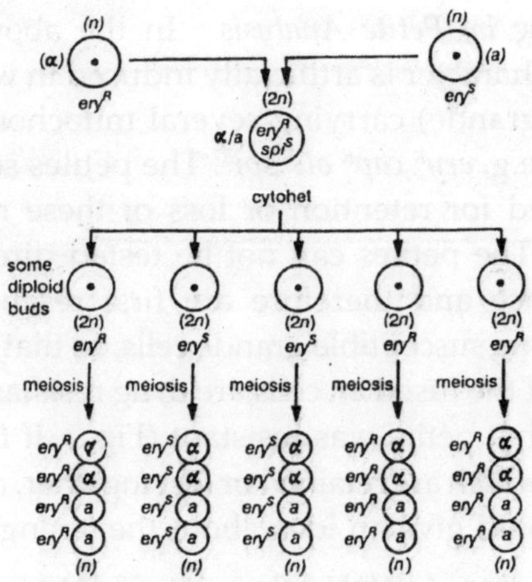

Non-Mendelian inheritance of erythromycin resistance cery) in yeast (alleles a and r α representing mating types segregate in a Mendelian fashion).

Recombination is also influenced by a genetic factor, omega (ω), not found in all yeast strains. Nevertheless recombination data like the above have been utilized for mapping, recombination generally never exceeding 25% as against 50% for nuclear genes.

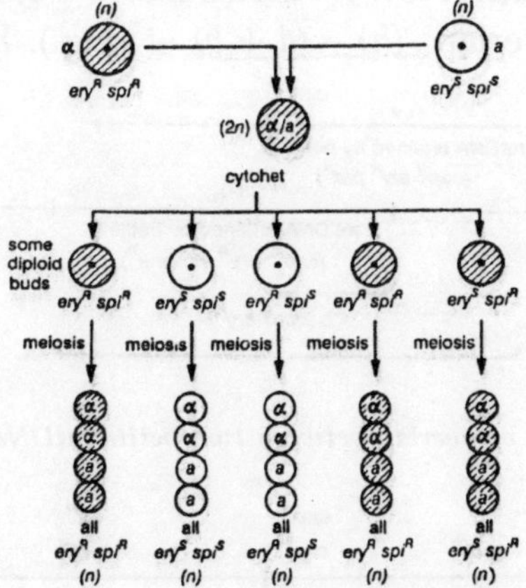

Non-Mendelian inheritance (from the cross ery^R spi^R × ery^S spi^S in yeast (spi^R = spiramycin resistance); note the cytoplasrhic segregation and recombination (CSAR) at the diploid level and lack of segregation during meiosis.

(b) *Mapping by Petite Analysis* : In the above technique, petite character is artificially induced in wild type (also called grande) carrying several mitochondrial marker genes, e.g. ery^R cap^R oli^R spi^R. The petites so induced are screened for retention or loss of these mitochondrial genes. The petites can not be tested directly for drug resistance and therefore are first rescued by fusion with drug susceptible grande cells, so that if on budding some of the fused 2n cells are drug resistant, this would mean that petite was resistant (Fig.). If frequencies of genes, which are retained or lost together, are compared, this would give an idea about the strength of linkage.

(c) *Overlapping of mtDNA Segments in Petites* : In the above analysis, petites can be identified, which retain different sets of mt-genes. For instance, suppose petite A retains cap^R *eryR parR* and petite B retains cap^R *eryR oliR anaR*. Now, DNA can be extracted separately from A and B, and hybridized with DNA from wild (grande) type cells. Suppose amount of DNA hybridized is *a* units with petite A, *b* units with petite B and *t* units with A + B together. Now if *t* is less than *a* + *b*, then there must be an overlap (*h*) – (*a* + *b*) –*t* (Fig.). In this overlap

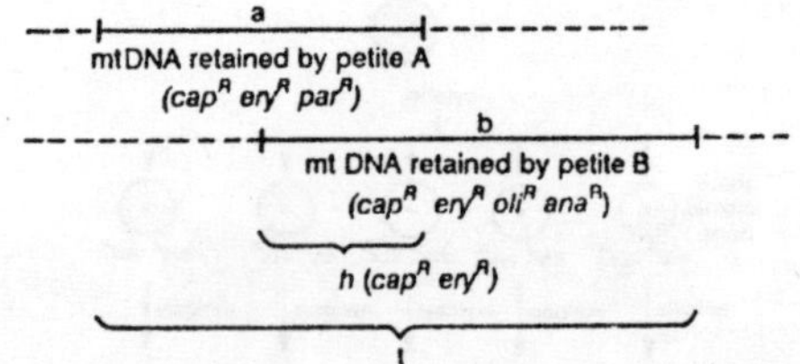

Measurement of overlap between two petite mtDNAs in yeast.

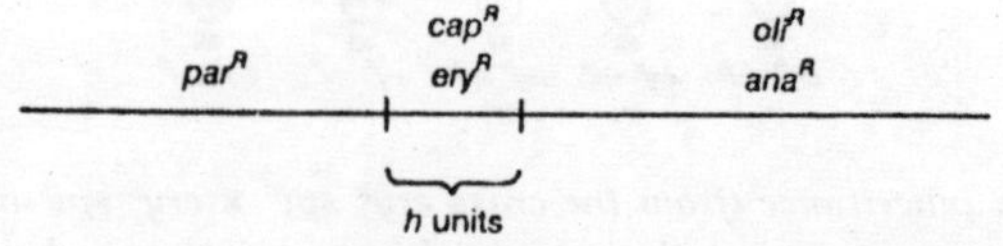

Placement of markers on genetic map using overlapping segments in two petite mtDNAs (shown in Figure) in yeast.

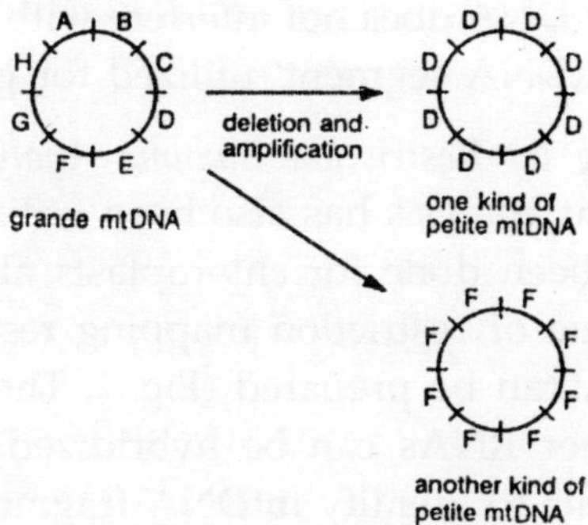

Deletion and amplification of mtDNA, leading to production of two different petites in yeast.

region *h*, must be found genes retained both in petite A and petite B. (cap^R ery^R in this case). The overlap *h* should also represent the maximum distance between cap^R and ery^R. This will allow placement of markers into map segments of defined size (Fig.). A library of petites is thus now available with various genes retained, and different petites related due to specific overlap regions. An interesting feature of petites is that the amount of mtDNA in petite and grande is almost equal. This is due to amplification of a small DNA segment which is retained in petite (Fig.).

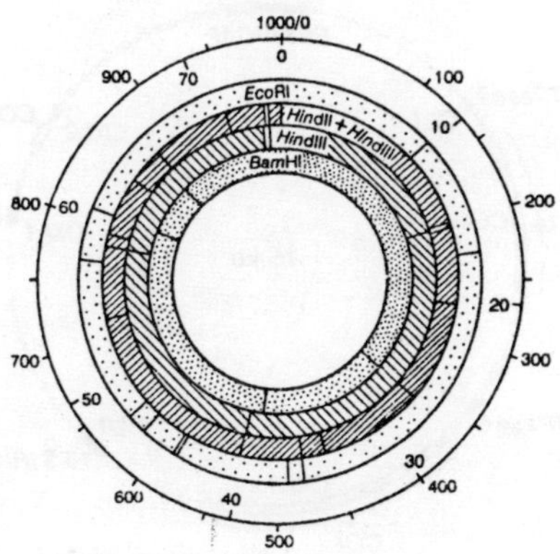

A restriction map of yeast mtDNA (figures in outermost circle represent kilobase pairs of DNA; inner four circle show restriction site for four enzyme combinations).

This equality of DNA does not interfere with the technique of 'overlapping mtDNA segment' utilized for genetic mapping.

(d) *Mapping by Restriction Enzyme Analysis* : Restriction fragment analysis has also been utilized for mapping as has been done for chloroplasts also. Utilizing the technique of restriction mapping restriction maps of mtDNA can be prepared (Fig.). The tRNAs, rRNAs and other RNAs can be hybridized with restriction fragments to identify mtDNA fragments carrying the genes. The limits of these fragments can be narrowed using fragments produced by different restriction enzymes separately, for hybridization with the same RNA.

Restriction mapping may also be combined with techniques of 'petite analysis' and 'overlapping DNA segments' described above. This allows us to locate physically the genes for drug resistance (*e.g.* ery^R, cap^R, etc.), on specific restriction fragments. The circular genetic map of yeast prepared following above techniques is presented in Figure. The genetic map of human mitochondria is show in Figure.

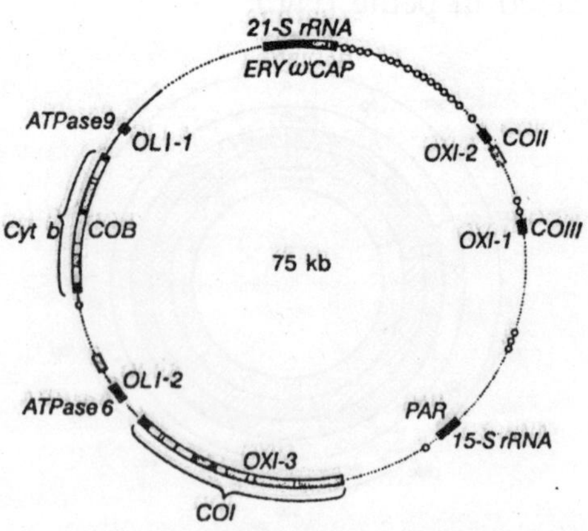

Circular genetic map of yeast (Saccharomyces cerevisiae). *Black boxes show regions of mRNA or rRNA synthesis. Small circles are regions of tRNA synthesis.*

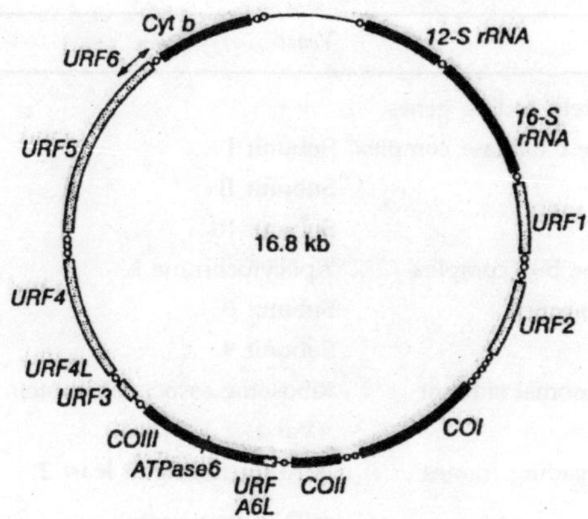

Map of human mitochondrial DNA.

Human mtDNA : In human beings, mt-DNA is smaller in size (16.8 kilobases) and the complete nucleotide sequence of this DNA is now known. From the sequence, without any genetic analysis, 13 genes for different prilypeptides, genes for two species of rRNA and those for a complete set of tRNA molecules could be identified. This is a remarkable example of complete genetic map obtained without any genetic analysis (Fig.).

Plant mtDNA : In maize, cytoplasmic male sterility is determined by mtDNA, which has already been discussed in detail earlier, while discussing male sterility in this chapter. The structure of mtDNA in several flowering plants is under investigation now. It seems that in plants, mtDNA molecules of variable size are available, making the study relatively difficult.

Genetic Content of Yeast and Mammalian mtDNA.

Class of genes	*Yeast*	*Mammals*
1. Ribosomal RNA genes	21S rRNA	16S rRNA
	15S rRNA	12SrRNA
2. tRNA genes	25 tRNA species	22 tRNA species

Contd...

Class of genes	*Yeast*	*Mammals*
3. Identified protein coding genes		
(i) Cytochrome c oxidase complex	Subunit I	—
	Subunit II	—
	Subunit III	—
(ii) Cytochrome b-c_1 complex	Apocytochrome *b*	Apocytochrome *b*
(iii) ATPase-complex	Subunit 6	Subunit 6
	Subunit 9	—
(iv) Small ribosomal subunit	Ribosome associated protein (Var-1)	?
4. Unidentified reading frames (URFs)	≥ 10 (including at least 2 mRNA maturases)	8

A Comparison of mtDNA in Yeast and Mammals : We have seen that the mitochondrial DNA of only a few individuals has been analysed in detail so far. In case of yeast and mammals, however, some clear picture has emerged. The sets of known genes in these two classes of organisms are identical except some differences. For instance, yeast has 25 tRNA species (more iso-accepting tRNAs), while mammals have only 22 tRNA species. Further, the genes for subunit 9 of the ATPase complex and those for a ribosome associated protein, though present in yeast, are absent in mammals. Mammalian mtDNA is transcribed into a single transcript from the major coding strand, and individual products are generated by RNA processing. In yeast, rearrangements can occur in mtDNA rather frequently.

A substantial fraction of the genetic material in mtDNA in yeast as well as in mammals is represented by unidentified reading frames (URFs). Most of the yeast URFs are contained in introns of split genes and appear to code for protein required for mtRNA splicing, the so called maturases. The URFs of mammalian mtDNA code for proteins of unknown function.

Unidentified Reading Frames (URFs) and Protein Coding Genes in mtDNA : In animals and particularly in mammals,

13 reading frames, coding for proteins (ranging from 70-610 amino acids) have been identified. These are in addition to the genes for ribosomal RNAs (rRNAs) and tRNAs. Five of these 13 reading frames code for cytochrome-c oxidase subunit I (COI), II (COII) and III (CO III), ATPase subunit 6 and cytochrome b. The remaining eight reading frames, representing 60% of the protein coding sequences, pertain to the so called URFs, whose products have not been identified with any protein of known function. Nothing is known about the physiological role of the products of mammalian mtDNA URFs, except that they are hydrophobic proteins associated with the inner mitochondrial membrane. URFA6L is perhaps homologous to *aapl* coding for a protein of 48 amino acids, which may be a part of ATPase complex.

Unidirectional Replication of mtDNA (D loops) : Another interesting feature of mtDNA is its unidirectional and highly asymmetric replication. The daughter L-strand starts synthesis when two third of H-strand is already synthesized. The position of origin of synthesis of H-strand is marked with displacement loop (D-loop). More details about D-loop and mtDNA synthesis.

Evolutionary Origin of Eukaryotic Mitochondria from Prokaryotes : A comparison between the properties of mitochondria and bacteria has shown similarities which indicate evolutionary relationship. Some of these similarities are listed in Table. On the basis of these similarities it has been suggested that mitochondria might have evolved due to symbiotic association of bacteria with cells of higher organisms.

In Table, the last three similarities between mitochondria and bacteria are rather very conspicuous. These include size and shape of DNA, there being circular DNA in both. The DNA in eukaryotic chromosomes is very different than circular DNA found in mitochondria in the same cells. Further, the protein synthesis in eukaryotic cytoplasm is inhibited by cycloheximide and not by chloramphenicol as in bacteria and mitochondria. This property has been utilized to stop preferentially the protein synthesis in cytoplasm by cycloheximide and thus mark the proteins synthesized in mitochondria or chloroplasts, both responding in the same manner to cycloheximide and chloramphenicol.

Similarly, protein synthesis in mitochondria and chloroplasts can be preferentially stopped by chloramphenicol and the proteins synthesized in cytoplasm can then be identified. This property of protein synthesis mechanism is due to similar ribosomes (70S) in chloroplasts, mitochondria and prokaryotes, because the ribosomes in cytoplasm of eukaryotes are very different (80S).

Comparison between Mammalian Mitochondria and Bacteria.

Characters	*Mitochondria*	*Bacteria*
1. Size: width	0.2-1.0 μ	0.2-2.0 μ
length	1.0-4.0 μ	0.3-10.0 μ
2. Lipoprotein membrane	60-70Å	70-80Å
3. Invaginations of membrane	Cristae	Mesosomes
4. Respiratory control	Marked	Low or absent
5. Inhibitors of phosphorylation like CN—, CO, azide, anti-mycin, DNP and oligomycin	Less effective Effective	or ineffective
6. DNA shape	Closed circle	Closed-circle
7. Inhibitor of protein synthesis	Chloramphenicol	Chloramphenicol
8. Ribosomes	60S-78S	70S

Paternal Inheritance of cpDNA and mtDNA : Since chloroplasts and mitochondria are organelles found in the cytoplasm, outside the nucleus, they are often transmitted with the cytoplasm. As discussed earlier, during sexual reproduction most of the cytoplasm in the zygote is derived from the egg (female parent), so that one expects that organellar

DNA is maternally transmitted. Besides the relatively predominant inputs of cpDNA and mtDNA from the female parent to the zygote, the inheritance of cpDNA and mtDNA is also influenced by vegetative sorting and by differences in rates of replication of organellar DNAs derived from the two parents. This leads to elimination of paternal cpDNA and paternal mtDNA leading to maternal inheritance.

For many years now, the cpDNA in angiosperms is known to be either maternally inherited or biparentally inherited. In contrast to this till recently, mtDNA in angiosperms and in animals was known to be strictly maternally inherited. Only in late 1980s and early 1990s, partly due to the use of backcrosses to male parent, for increasing the concentration and partly due to the availability of RFLP and PCR techniques, examples of paternal and biparental inheritance of mtDNA both in plants and animals became available. For cpDNA also, several studies in conifers (Gymnosperms), conducted during late 1980s and early 1990s, revealed strictly paternal inheritance, only few offspring having maternal or novel genotypes.

Paternal transmission of cpDNA and mtDNA, using specific molecular makers, has been shown for many cases including the following :

(i) In barley-rye intergeneric crosses, some paternal inheritance of mtDNA was observed,

(ii) In alfalfa, biparental inheritance of mtRNA was observed,

(iii) In intraspecific crosses of Douglas fir *(Pseudotsuga menziesii)*, as well as in interspecific crosses in the genera, *Pinus, Larix* and *Picea*, strictly paternal inheritance of cpDNA was observed,

(iv) In two intraspecific crosses of redwood *(Sequoia sempervirens)*, both cpDNA and mtDNA were strictly paternally inherited (as shown by using specific chloroplast DNA probes derived from *Petunia* and mitochondrial DNA probes derived from maize),

(v) In an interspecific cross in mice *(Mus domesticus* x *M. spretus),* where concentration of paternal mtDNA was increased by backcrossing, paternal mtDNA was transmitted at a rate of 10 relative to maternal mtDNA (as shown by amplification of paternal mtDNA using polymerase chain reaction),

(vi) In hybrid crosses in *Drosophila* also, incidental paternal transmission of mtDNA was reported,

(vii) In 1991, interspecific crosses *(Mytilus edulis* x *M. trossulus)* and intraspecific crosses in mussels (a mollusc, like *Unio)* were used to demonstrate extensive contribution of paternal mtDNA.

Genetics of Cancer

Cancer cells are those cells which have lost the usual 'control over their growth and division. They are characterized by immortalization (indefinite growth), transformation and metastasis (invasion of normal tissue). Metastasis really distinguishes a malignant tumour (ability to invade normal tissue) from a clinically benign (does not invade new tissue) tissue. In culture also, cancerous cells differ from the established cell lines, which are characterized by the dependence of their growth on the availability of a 'limited density' (growth is inhibited beyond this density).

The cancerous cells do not need a solid surface for growth, have reduced dependence on serum, may form a thick mass of cells (instead of a thin layer) and induce tumours when injected into appropriate test animals. These cancer cells may be compared with normal cells to identify the genetic basis of tumour formation.

Tumour formation may be spontaneous, or induced by carcinogens or tumour viruses. This is often caused due to accumulation of mutations, which promote clonal selection of cells that have lost control over their growth and division. Vast majority of these mutations are somatic and are found only in an individual's cancer cells.

However, about 1% of all cancers arise in individuals with an unmistakable hereditary cancer syndrome. These individuals carry a particular germline mutation in every cell of the body. A number of genes have now been identified in viruses and in the host cells that have the potential of causing tumour formation. They are called oncogenes (roughly five dozen oncogenes were already discovered by the end of 1990) and are described as viral or *v-onc* genes (present in viruses) and cellular or *c-onc* genes (present in host cells).

Often the genes present in host are not oncogenic and are, therefore, described as proto-oncogenes, which have the potential of becoming oncogenes. There are also genes which suppress tumour formation. These are called anti-oncogenes or tumour suppressor genes.

A variety of these genes are now known and will be discussed in this chapter.

Viral Oncogenes (v-onc Genes) : Certain tumour viruses (including both DNA and RNA viruses) carry genes, which

Transforming Viruses may Carry Oncogenes

Viral class	*Type of virus*	*Oncogenes*	*Origin of oncogenes*
Papova	DNA duplex	T antigens	Early viral gene
Adeno	DNA duplex	E1A & E1B	Early viral gene
Epstein-Barr	DNA duplex	BNLF-1	Latent viral gene
Retrovirus (acute)	RNA (single stranded)	Individual	Cellular
Retro (non-defective)	RNA (single stranded)	None	—

confer on them the ability to convert host cells to a tumorigenic state. These genes are called oncogenes. The DNA and RNA viruses differ in some important characteristics which will be briefly described:

(i) DNA viruses. The DNA tumour viruses may also grow in culture on a variety of cells, which can be classified into following two types : (a) 'permissive cells', which can be productively infected and (b) 'non-permissive cells', which are not productively infected, but merely get transformed. Most oncogenes in these viruses are expressed by alternative splicing. Unlike RNA viruses, oncogenes in these DNA viruses do not have cellular counterparts (see later).

(ii) RNA viruses. RNA tumour viruses may be transformation competent (able to transform the infected cells) or transformation-incompetent (able to replicate, but not transform the infected cells). They may replicate via DNA intermediate or an RNA intermediate and can transfer genetic information horizontally or vertically. The RNA viruses may be classified in two groups with respect to their tumorigenicity. (a) Non-defective viruses. These viruses cause tumour formation, not due to an individual oncogene, but due to its ability to activate a cellular gene(s); two classic models are FeLV (Feline leukaemia virus; Feline = of cats) and MMTV (mouse mammary tumour virus). (b) Acute transforming viruses. These viruses, each possesses an oncogene (absent in the ancestral virus, but derived later from host cell due to a transducing agent), which induces tumour formation *in vivo* rather rapidly, and transforms cultured cells *in vitro*. These transforming viruses are classified further according to the type of tumour, which is caused (*e.g.* leukaemia, sarcoma, carcinoma, etc.) and the type of tumour is dependent on the oncogene, a virus carries.

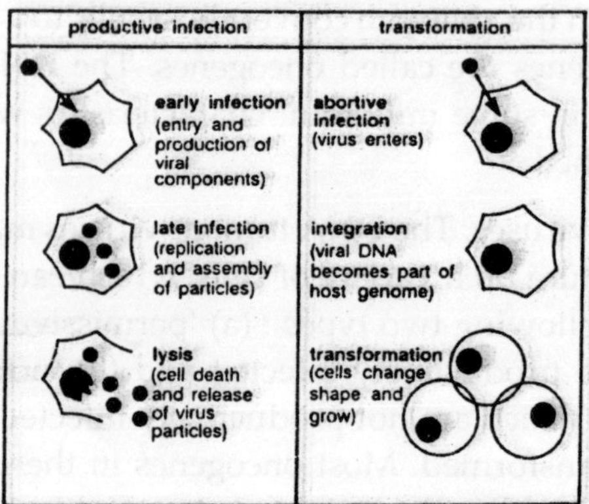

Response of permissive cells and non-permissive cells to virus infection; permissive cells are productively infected by DNA tumour virus, which leads to lysis, but non-permissive cells get transformed through change in their phenotype.

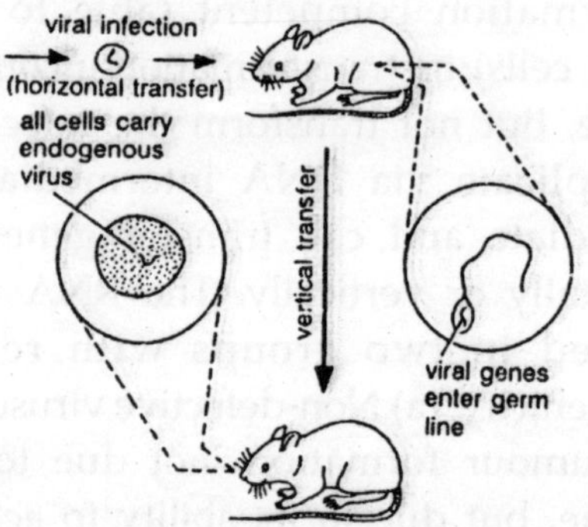

Horizontal and vertical transfer of genetic information from retroviruses; horizontal transfer involves infection of new host, while vertical transfer involves transfer from one generation to the next generation, after integration of viral genome in the germ line of the host.

Oncogenes of some retroviruses are listed in Table. These oncogenes are related to some sequences in the cellular genome, called proto-oncogenes, which are not oncogenic, but can give rise to an oncogene, particularly when captured by a retrovirus. The viral oncogenes and cellular proto-oncogenes are distinguished by using prefixes v and *c* so that an oncogene carried by Rous Sarcoma Virus is called *v-src* and the related

Each Transforming Retrovirus Carries an Oncogene Derived from a Cellular Gene

Virus (name in parentheses)	*Species*	*Tumor*	*Oncogene*
Rous sarcoma (RSV)	Chicken	Sarcoma	*sre*
Harvey murine sarcoma (Ha-MuSV)	Rat	Sarcoma and erythroleukaemia	*H-ras*
Kirsten murine sarcoma (Ki-MuSV)	Rat	Sarcoma and erythroleukaemia	*K-ras*
Moloney murine sarcoma (Mo-MuSV)	Mouse	Sarcoma	*mos*
FBJ murine osteosarcoma (FBJ-MuSV)	Mouse	Chondrosarcoma	*fas*
Simian sarcoma (SSV)	Monkey	Sarcoma	*sis*
Feline sarcoma (Pl-FeSV)	Cat	Sarcoma	*sis*
Feline sarcoma (SM-FeSV)	Cat	Fibrosarcoma	*fms*
Feline sarcoma (ST-FeSV)	Cat	Fibrosarcoma	*fes*
Avian sarcoma (ASV-17)	Chicken	Fibrosarcoma	*jun*
Fujinami sarcoma (FuSV)	Chicken	Sarcoma	*fps*
Avian myelocytomatosis (MC29)	Chicken	Carcinoma, sarcoma, and myelocytoma	*myc*
Abelson leukaemia (MuLV)	Mouse	B cell lymphoma	*abl*
Reticuloendotheliosis (REV-T)	Turkey	Lymphatic leukaemia	*rel*
Avian erythroblastosis (AEV)	Chicken	Erythroleukaemia and fibrosarcoma	*erbB (erbA?)*
Avian myeloblastosis (AMVO	Chicken	Myeloblastic leukaemia	*myb*

proto-oncogene in the cellular genome is called *c-src*. *v-onc* is actually derived from m-RNA, so that it contains uninterrupted sequence (or sequence missing at one or both ends), while the corresponding *c-onc* may be a split gene with exons and introns. The *mos, ras, sis* and *myc* genes have been derived from entire *c-onc* genes having no missing regions (Table). A *v-onc* sequence is often expressed as a part of a fusion product, where the boundaries can be defined by comparing with the corresponding *c-onc*. Several such examples include gag-abl (from *v-abl*), gag-fps (from *v-fps*), gag-myc (from *v-myc*) and gag-fos (from *v-fos*) (see Table).

Oncogenicity may be Caused by Quantitative or Qualitative change of c-onc _Genes_

Onco protein	*Onco-genicity of excess c-onc*	*Oncogenic sequence changes*
src	None	Substitution at C-terminus and other sequence changes
abl	Weak?	Linkage to gag sequence in *v-abl* or N-terminal substitution in *c-abl*
erbB	Weak	Deletion of extracellular (ligand-binding) domain & loss of C-terminus
fms	None	Point mutation in extracellular (ligand-binding) domain (N-terminal)
neu	?	Point mutation in transmembrane domain
mos	Weak	C-terminal sequence of *c-mos*
ras	Weak	Changes at codons 12, 59, 61
myc	Effective	Not necessary

Atleast 20 *c-onc* genes have been identified by their representation in retroviruses. Same *c-onc* gene may be found in more than one retroviruses and more than one *c-onc* gene may be present in different strains of the same retrovirus, although the different *v-onc* genes may also be related (derived from different members of same *c-onc* gene family (*e.g. c-ras* genes or *c-myc* genes). A wide variety of mechanisms that activate proto-oncogenes *(c-onc)* are listed in Table.

Oncoproteins have a Variety of Locations

Gene	*c-onc protein location*	*v-onc protein*	
		Compo-nents	*Location*
src	Membranes	src	Plasma & nuclear membranes
abl	Nucleus	gag-abl	Cytoplasm
erbB	Plasma membrane	P68-erbB	Plasma membrane & golgi
neu	Plasma membrane	—	—
fps	soluble	gag-fps	Soluble & membrane
fms	Plasma membrane	fms	Plasma membrane
mos	Cytoplasm	env-mos	Cytoplasm
ras	Membranes	ras	Membranes
myc	Nucleus	myc	Nucleus
	Nucleus	gag-myc	Nucleus
myb	Nucleus	myb	Nucleus
fos	Nucleus	gag-fos	Nucleus

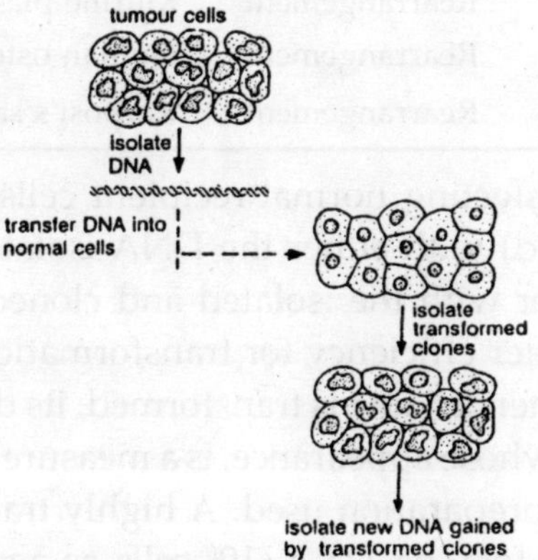

Different steps of a transfection assay.

That the *v-onc* sequence is actually responsible for oncogenic potential of a transforming virus, has been proved by using following two approaches : (i) *v-onc* genes could be mutated to abolish its transforming activity; (ii) *c-onc* proto-oncogenes can be altered to become oncogenic. Changes in protein sequence of the product derived from *c-onc,* relative to that derived from *v-onc* is not always necessary for oncogenicity. In some cases, mere over-expression or altered regulation may lead to oncogenic phenotype.

Isolated Oncogenes for Transfection Assay

Viral infection is not really necessary for tumour formation. We can actually assay for transformation

Several Oncogenes can be Detected by the Ability to Transform Mouse 3T3 Fibroblasts in a Transfection Assay

c-onc	*v-onc*	*Change in c-onc*	*Type of tumour carrying oncogene*
H-ras	*H-ras*	Point mutation	—
K ras	*K-ras*	Point mutation	Various (human); see Table.
N-ras	none	Point mutation	—
neu	none	Point mutation	Rat neuroblastoma
mos	*mos*	Rearrangement	Murine plasmacytoma
met	none	Rearrangement	Human osteosarcoma
KS	none	Rearrangement	Karposi's sarcoma

by directly transfecting normal recipient cells (3T3 fibroblast line is often used) with either the DNA isolated from animal tumours (Fig.) or with the isolated and cloned genes (cloned genes have greater efficiency for transformation). When a cell in 3T3 line or other cell line is transformed, its descendants pile up into a focus, whose appearance, is a measure of transforming ability of DNA preparation used. A highly transforming gene may give > 100 foci/ng DNA/10^6 cells as against < 10 foci/ng DNA/10^6 cells obtained with a weakly transforming gene.

A number of isolated genes used for transformation are listed in Table.

Activation of Proto-oncogenes by Mutations

Ras Proto-oncogenes : There are three active genes (N, H and K) belonging to the *c-ras* family (found both in man and rat), which give rise to oncogenes. All three genes are related with conserved sequences and some variable regions. The *v-H-ras* and *v-K-ras* differ from corresponding *c-ras* genes by alterations of 3 and 7 amino acids respectively. It has also been shown that each *c-ras* proto-oncogene can give rise to an oncogene (found in human tumours) by a single base replacement leading to the substitution of a single amino acid at position 12 or 61 in one of the Ras proteins (Table). It has also been shown that the DNA from a lung carcinoma has a

The* Ras *Gene Family Provides Oncogenes in Human Tumours and also is Carried by Transforming Retroviruses.

Proto-oncogene	*Aminoacid at*			*Source of tumor*
	12	*59*	*61*	
c-ras (H,K,N)	Gly	Ala	Gin	Wild type sequence
H-ras	+	+	Leu	Lung carcinoma Hs242
	Val	+	+	Bladder carcinoma T24
K-ras	Cys	+	+	Lung carcinoma Calu-1
	Arg	+	+	Lung carcinoma LC-10
	Val	+	+	Colon carcinoma SW480
N-ras	+	+	Lys	Neuroblastoma
	+	+		SK-N-SH
			Arg	Lung carcinoma
				SW1271
				Murine Sarcoma Virus
H-ras	Arg	Thr	+	Harvey strain
K-ras	Ser	Thr	+	Kirsten strain

mutant *c-ras* gene, while DNA from normal tissue of same cancer patient has a wild type *c-ras* gene. Such observations suggest that a spontaneous mutation in *c-ras* gene may occur in any tissue leading to tumour formation. The oncogenicity of *c-ras* genes may also be enhanced by over-expression or amplification of *c-ras* genes, suggesting that oncogenesis depends on over-activity of Ras protein.

Activation of Proto-oncogenes by Insertions, Translocations and Amplification : There are examples available, when events like insertions, translocations or amplification of DNA sequences of proto-oncogenes lead to their activation, although the nucleotide sequence remains unaltered. These mechanisms have been studied in *c-myc,* whose expression is elevated by these events leading to tumour formation.

Insertions of Retroviral Sequences : The *c-myc* gene has three exons (El, E2, E3), El representing the first untranslated (but transcribed) leader sequence, and the remaining two coding for *c-myc* protein. Following three classes of insertions in *c-myc* gene are known (Fig.). (i) The retroviral genome may be inserted in the first intron, so that its LTR (long terminal repeat) works as an efficient promoter increasing the level of expression of *c-myc* protein coded by El and E2. The transcript also lacks the untranslated leader sequence. (ii) The retroviral genome is inserted in reverse orientation in the first intron or upstream to this intron and its LTR works as an enhancer. (iii) The retroviral genome may also be inserted downstream of *c-myc* gene, and LTR may work as an enhancer.

There are also other oncogenes, which are activated to make tumours by the insertion of retroviral genome. Following are some examples : *c-erbB, c-myb, c-mos, c-H-ras* and *c-raf, int 1, int* 2.

Chromosomal Translocations : In some cancerous cells, chromosomal translocations have been discovered, where an oncogene of one chromosome is brought into the proximity of an *Ig* locus on another chromosome (*e.g.* plasmacytomas in the

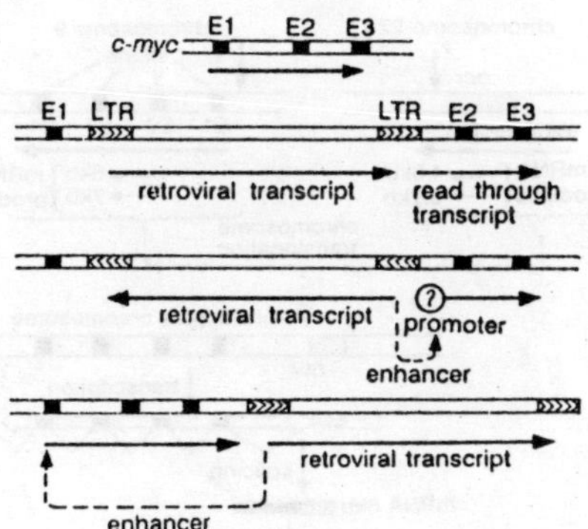

Three different kinds of known insertions of retroviral genome in* c-myc *gene, leading to activation of* c-myc *gene.

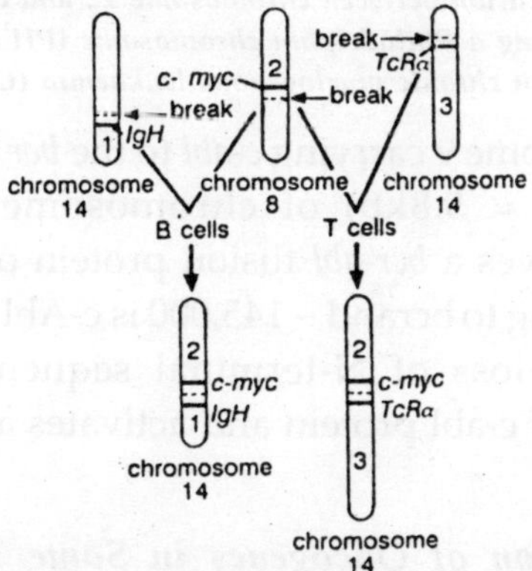

Translocations leading to activation of* c-myc *gene; in B cells* c-myc *is translocated to the proximity of* Ig *locus, while in T cells, it is translocated to the vicinity of* TcRa *locus (in both cases* c-myc *gets activated).

mouse and Burkitt lymphomas in man). Such transiocations lead Jo activation of a proto-oncogene like *c-myc* (Fig.), which is located on chromosome 8 in humans and on chromosome 15 in mouse. Due to translocation, *c-myc* is brought in close proximity of *Ig* gene on chromosome 14 in humans and chromosome 12 in mouse. Another case, where a translocation activates an oncogene involves Philadelphia (PH^1) chromosome found in chronic myelogenous leukaemia (CML). This translocation involves transfer of a 5000kb region from the

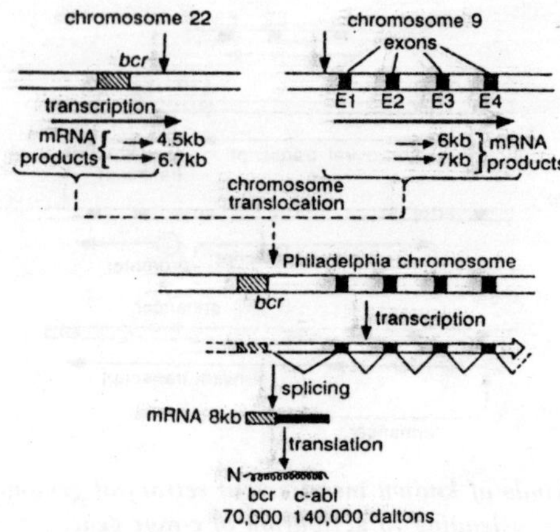

A translocation between chromosome 22 and chromosome 9 generating a Philadelphia chromosome (PH[1]), associated with chronic myelogenous leukaemia (CML).

end of chromosome 9 carrying *c-abl* to the *bcr* locus (breakpoint cluster region = 5.8kb) of chromosome 22 (Fig.). This translocation gives a *bcr-abl* fusion protein of 210,000 daltons (70,000 belonging to bcrand ~ 145,000 is c-Abl protein). Perhaps the fusion, or loss of N-terminal sequence, changes the conformation of c-abl protein and activates a latent oncogenic potential.

Amplification of Oncogenes in Some Tumours : Many tumour cell lines have visible regions of chromosomal amplification containing a known oncogene. In other cases a particular oncogene is amplified, as revealed by molecular probes, although the amplification is not visible from chromosome morphology.

This amplification leads to over-expression of the gene leading to tumour formation. In amplified DNA only, two new cellular genes *N-myc* and *L-myc* were discovered, which had partial homology with *c-myc*.

Transgenic Mice for a Study of Oncogenic Potential of c-myc Gene : A normal *c-myc* gene linked to either an Ig-H enhancer or to an LTR (enhancer) derived from MMTV, when used for the production of transgenic mice, the latter developed a variety of cancers. In these cases over-expression of *c-myc* gene is responsible for the development of cancers.

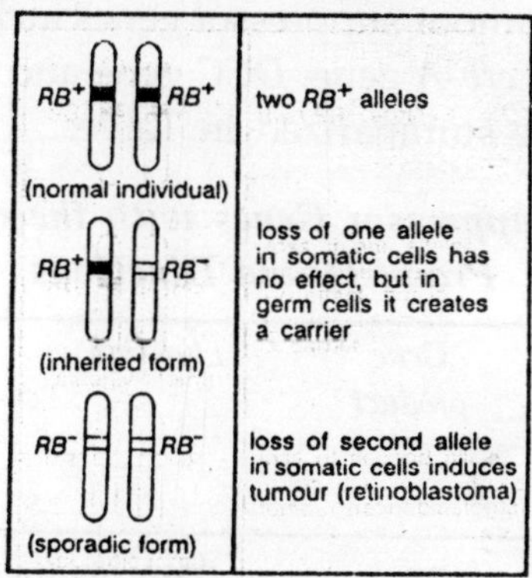

The effect of loss of one or both RB *genes involving 13q14 region in humans.*

Tumour Suppressor Genes or Anti-oncogenes : The formation of cancerous tissues is regulated by growth promoting proto-oncogenes, which are counterbalanced by growth suppressing genes called 'tumour suppressor genes'. The growth lesions, that inactivate suppressor genes, liberate the cells from the constraints imposed by these genes, yielding the unconstrained growth of the cancer cells. It is possible that activation of proto-oncogenes and inactivation of tumour suppressor genes, both are responsible for the progression of many tumours to full malignancy. Many of these tumour suppressor genes have been isolated as molecular clones and have been subjected to intensive investigation.

The evidence about the presence of tumour suppressor genes came from the following types of experiments : (i) Somatic cell hybridization, where fusion of tumour cells with normal cells always gave non-tumorigenic hybrid cells, suggesting that the normal cells donated the tumour suppressor genes, (ii) In patients suffering with retinoblastoma (RB), tumour on the retina in children, was associated with deletion of band q14 of chromosome 13. *RB* gene was later associated with 13ql4, and retinoblastoma involved inactivation or deletion of both copies of *RB* gene.

Some of the tumour suppressor genes include *RB* gene, *p53* gene, *WT-1* gene, *erb-A* gene, *DCC* gene and *NF-1* gene. Their characteristics are summarized in Table.

Some Tumour Suppressor Genes with their Location, Gene Products and Functions

Gene	*Location on chromosome*	*Gene product*	*Function*
RB	13ql4	pRB (1055 kD)	DNA binding ability; inactivated by forming complexes with E1A; T antigen cell cycle regulation by E7; reversible phosphorylation; interacts with transcription factors.
p53	17p	p53	Suppresses effect of *ras;* forms complex with hsc70
WT-1	11pl3	WT-1 (345 amino acids)	Very much like transcription factor (four zinc-finger domains)
DCC	18q	Phosphoprotein (190 kD)	—
NF-1	17q11.2	NF-1	Stimulates GTPase activity; down regulator of $p21^{ras}$

Inherited Cancer Genes

As discussed above, a vast majority of cancers though genetic in nature, are caused by somatic mutations only, and therefore are not transmitted to the next generation. In other words, they do not reach the germline. However, 1% of all cancers are due to mutations located in the germline and are therefore inherited. The first such inherited cancer gene *(RBI)* was cloned in late 1980s and during a period of about 10 years (up to 1997), more than 20 different hereditary cancer syndromes were defined and attributed to specific germline mutations in more than 25 different inherited cancer genes (Table). About 80% of these inherited cancers are dominant in nature, and are often due to mutations in tumour suppressor genes.

However, even when a germline mutation in a cancer gene is inherited, it is not always necessary, that an individual carrying this gene should suffer with cancer. The individual, who carries a mutant allele of an inherited cancer gene has a variable risk of cancer, which is influenced by a number of factors including the mutation itself, other cellular genes, diet, lifestyle and environmental factors. For instance, there are modifier genes, which play a significant role in cancer development. Therefore, inherited cancer

A Summary of Selected Inherited Cancer Syndromes

Syndrome	*Cloned gene*
I. Dominant Inheritance	
1. Familial retinoblastoma	*RB1*
2. Li-Fraumeni syndrome	*p53*
3. Familial adenomatous polyposis (FAP)	*APC*
4. Hereditary nonpolyposis colorectal cancer (HNPCC)	*PMS1, PMS2 MSH2, MLH1*
5. Neurofibromatosis type 1 and type 2(NF1,NF2)	*NF1.NF2*
6. Wilms tumour	*WTI*
7. Weidmann-Beckith syndrome (WBS)	*p57/KIP2*
8. Nevoid basal cell carcinoma syndrome (NBCCS) .	*PTCH*

Contd...

Syndrome	Cloned gene
9. Familial breast cancer, 1, 2	*BRCA1 BRCA2*
10. Von Hippel Lindau (VHL) syndrome	*VHL*
11. Hereditary papillary renal cancer (HPRC)	*MET*
12. Familial melanoma	*p16 (CDKN2)*
13. Multiple endocrine neoplasia type 1 and type 2 (MEN1, MEN2)	*MENI, RETVHL*
14. Multiple exostoses	*EXT1, EXT2, EXT3*
15. Cowden disease	*PTEN (MMAC1)*
16. Hereditary prostate cancer (HPC)	*unknown*
II. Recessive Inheritance	
17. Atria telangiectasia (AT)	*ATM*
18. Bloom's syndrome (BS)	*BLM*
19. Xerodema pigmentosum (XP)	*XPB. XPD, XPA*
20. Fanconi's anemia (FA)	*FACC, FACA*

genes constitute only a subset of a larger class of genes that affect the cancer risk of an individual. This larger class of genes is sometimes termed as cancer susceptibility genes. A summary of some of the inherited cancer syndromes is presented in Table.

The proteins encoded by the above inherited cancer genes take part in diverse cellular processes including cell proliferation, differentiation, apoptosis (programmed cell death) and maintenance of genomic integrity. In performing the above functions, these proteins may function as (i) transmembrane receptors (MET, PTCH, RET); (ii) cytoplasmic regulatory or structural proteins (NF1, PTEN, APC, NF2); (iii) transcription factors or regulators of transcription (p53, WTI, RBI, VHL); (iv) cell cycle factors (CDK4 and pl6) or (v) proteins for DNA repair (MSH2, MLH1, PMS2, ATM, BRCA1, BRCA2, FACC, FACA,

XPA, XPB, XPD, BLM). It has also been shown that most of the inherited cancer genes exhibit a dominant pattern of inheritance and only some are recessive (Table). Further, most of these mutations lead to inactivation of tumour suppressor genes, rather than activation of oncogenes. Only three syndromes have been attributed to germline mutations in oncogenes *(RET, MET, CDK4)*. However, besides mutations in tumour suppressor genes or oncogenes, additional somatic mutations are needed for cancer development.

Cellular Functions of Oncoproteins

Oncoproteins and their corresponding proto-oncoproteins are variously distributed in the nucleus, cytoplasm and the nuclear membrane. On the nuclear membrane, they may be found on the cytoplasmic face or may be found as transmembrane proteins. Distribution of a number of oncoproteins is shown in Figure. They are also classified, on the basis of function into following four groups : (i) tyrosine kinases (subdivided into cytoplasmic *proteins* and membrane-bound growth factor receptors), (ii) growth factors, (iii) GTP-binding proteins and (iv) nuclear proteins.

Src (associated with RSV) was the first oncoprotein of the kinase type, that was characterized. Src is located on the cytoplasmic face of the cell membrane (Fig.). In the cells transformed by RSV, phosphotyrosine is increased ten times, and Src is responsible not only for its own phosphorylation (at residues 416, 527) but also for phosphorylation of tyrosine in several other proteins. Its catalytic activity resides in the c-terminal half of the protein and this activity in v-Src is 20 times that in c-Src. The oncogenic property of Src also depends on the replacement of N-terminal amino acid of Src by an unusual amino acid 'myristic acid', a phenomenon described as myristylation. How does phosphorylation of c-Src and v-Src regulates cellular function is being studied in some detail.

There are also proto-oncogenes, which code for receptors, residing as transmembrane domains. A model for the function of these receptors assumes that binding of a ligand to the extracellular domain activates the tyrosine kinase activity of

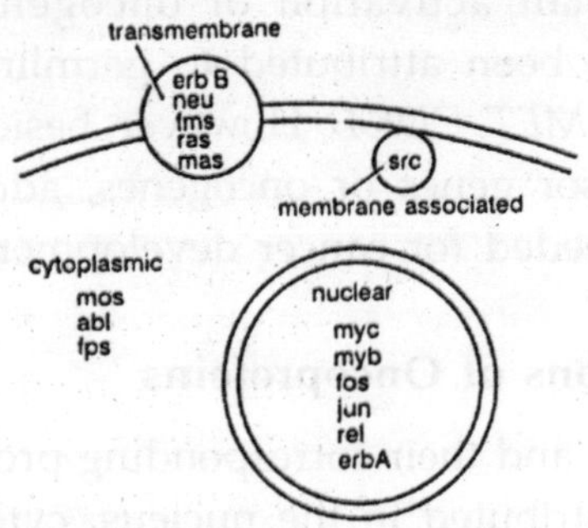

Location of oncoproteins—transmembrane proteins, membrane associated proteins, cytoplasmic proteins and nuclear proteins.

the intracellular domain, which leads to phosphorylation of cellular substrates. The best examples are EGF (epidermal growth factor) receptor coded by *c-erbB,* and CSF-1 (colony stimulating factor) coded by *c-fins.* There are also oncoproteins like pDGF (platelet derived growth factor), which stimulates the expression of several genes, including *c-myc* and *c-fos.*

Regulation of Gene Expression by Oncoproteins : There are several examples, where oncoproteins regulate gene expression : (i) *c-erb A* oncogerie of AEV codes for a protein that resembles a steroid receptor and may thus regulate gene expression; (ii) *c-jun* gene related to *v-jun* of ASV-17, codes for the transcription factor API or a part of it. AP-1 acts at enhancers and stimulates their activity. Actually fos and jun proteins form a complex that binds to DNA at the TGACTCA box and work together as a transcription factor of some significance.

E1A is another oncoprotein (phosphoprotein located in the nucleus), which regulates gene expression by interacting with several other cellular proteins. This protein activates expression of some genes and represses the activity of some other genes. It is capable of giving rise to three transcripts due to alternative splicing of the primary transcript.

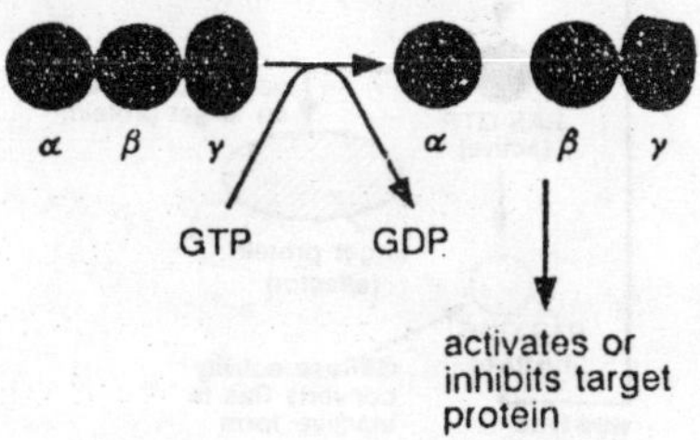

A typical trimeric (αβγ) G protein, where GTP binding releases α subunit from βγ dimer; GTP-α subunit or Py dimer becomes free to βγ upon target proteins.

Signal Transduction by Oncoproteins (G Proteins) : The concept of signal transduction was discussed and the most popular example involves G proteins, so named due to their ability to bind guanine residues. They are heterotrimeric proteins (with three subunits α, β and γ), located in plasma membrane and transduce extracellular signals received by transmembrane receptors to effector proteins in many eukaryotes (*e.g.*, yeast, plants, *Dictyostelium* and animals).

Since G proteins tranduce signals, they are also described as 'second messengers'. The intra-cellular effector molecules can be diverse, including adenyl cyclase, phospholipases and ion channels. Multiple genes encoding α, β and γ subunits of G proteins have been found in several organisms, with greatest sequence diversity discovered among the α subunits.

G proteins are activated, when GTP binds to the α subunit (by displacing GDP) and causes its dissociation from the βγ dimer. Such an activation- is common in a variety of GTP binding proteins. The separated βγ dimer may carry the message from receptor to effector, although in some cases a subunit may do this job.

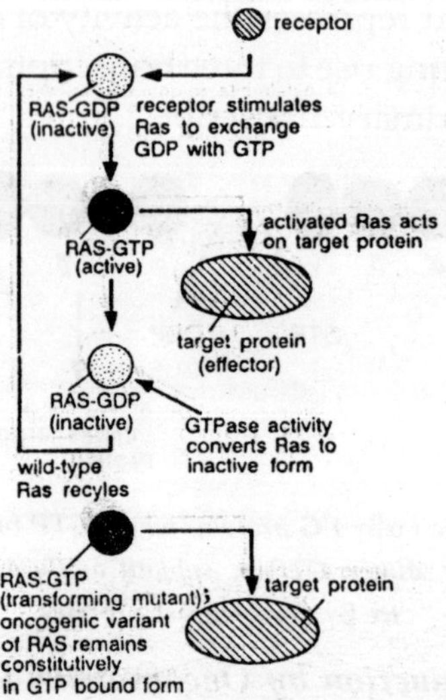

Activation of Ras, which stimulates replacement of GDP by GTP; the active protein recognizes its effector and GTP is cleaved into GDP, so that resulting inactive Ras is recycled; transforming Ras mutants do not hydrolyze GTP, so that Ras remains permanently in the active form.

Among oncoproteins, G proteins are represented by Ras proteins, which bind GTP and thus resemble α subunit. While bound to GTP, Ras becomes active and acts upon its target molecule. Following this interaction, GTPase activity of Ras protein hydrolyses GTP molecule to GDP, thus returning itself to inactive condition (Fig.).

Bibliography

Ahmad, K. Henikoff, S. : *Epigenetic Consequences of Nucleosome Dynamics*, MIT Press, Cambridge, 2002.

Altaian, S. : *Transfer RNA*, MIT Press, Cambridge, 1978.

Anderson, S. : *Sequence and Organization of the Human Mitochondrial Genome Nature*, Oxford University Press, London, 1981.

Andre, C. A. Levy and V. Walbot : *Structure of Plant Mitochondrial Genomes*, John Wiley, London, 1992.

Atkins, J. F. : *Recoding : Dynamic Reprogramming of Translation*, Oxford, New York, 1996.

Bennett, R. H. : *Transcription, Processing and Editing in Plant Mitochondria*, Oxford, New York, 1992.

Bloom, L. B. : *Biochemical Basis of DNA Replication Fidelity*, MacMillan, London, 1993.

Branden, C. and J. Tooze : *Introduction to Protein Structure*, Garland, New York, 1991.

Darnell, J. E., H. F. Lodish and D. Baltimore : *Molecular Cell Biology*, Scientific American Books, New York, 1995.

Das, A. : *Control of Transcription Termination*, Routledge, New York, 1993.

Davidson. J. N. : *Biochemistry of Nucleic Acids*, Cox and Wyman, London, 1965.

Demidov, V. V. : *PNA and LNA throw Light on DNA, Trends,* Oxford, New York, 2003.

DuPraw, E. J. : *Cell and Molecular Biology,* Academic Press, New York, 1969.

Ebby, S. R. : *Non Coding RNA Genes and Modern RNA,* Oxford, New York, 2001

Echols, H. : *Fidelity Mechanisms in DNA Replication,* Oxford, New York, 1991.

Ellis, R. J. : *Chloroplast Protein Synthesis,* Oxford, New York, 1975.

——— : *Molecular Chaperones,* Oxford, New York, 1991.

——— : *The Biogenesis of Chloroplasts Protein,* Oxford, New York, 1974.

Fischer, L.M. : *DNA Unwinding in Transcription and Recombination,* Oxford, New York, 1982.

Friedberg, E. C. : *Relationships Between DNA Repair and Transcription,* Oxford, New York, 1996.

Gesteland, R. F. and J. F. Atkins : *The RNA World,* Oxford, New York, 1993.

Gilbert, W. and B. Muller : *Isolation of the Lac Repressor,* Oxford, New York, 1966.

Gilham, N. W. : *Genetic Analysis of the Chloroplast and Mitochondrial Genomes,* Oxford, New York, 1974.

Gilmour, R. S. and J. Paul : *Tissue Specific Transcription of the Globin Gene in Isolated Chromatin,* Oxford, New York, 1973.

Glick, B. and G. Schatz : *Import of Proteins into Mitochondria,* Oxford, New York, 1991.

Gray, M. W. : *The Evolutionary Origin of Organelles,* Oxford, New York, 1989.

Grunstein, M. : *Histones as Regulators of Genes,* Oxford, New York, 1992.

Gupta, P. K. : *Cell and Molecular Biology,* Rastogi Publications, Meerut, 2000.

Hadjiolov, A. A. : *The Nucleolus and Ribosome Biogenesis,* Oxford, New York, 1985.

Hoagland, M. G. : *The Present Status of the Adaptor Hypothesis,* MacMillan, London, 1959.

Hoeijmakers, J. H. : *Nucleotide Excision Repair from Yeast to Mammals,* MacMillan, London, 2001.

Holley, R. W. : *Structure of a Ribonucleic Acid,* McGraw Hill, London, 2002.

Ingram, V. : *The Biosynthesis of Macromolecules,* Benjamin, New York, 2001.

Jacob, F. : *Genetic Regulatory Mechanisms in the Synthesis of Proteins,* Oxford, New York, 1961.

Jukes, T. H. : *Mitochondrial Codes and Evolution,* Oxford, New York, 2000.

Kaguni, L. S. : *Eukaryotic DNA Polymerase-primase : Structure, Mechanism and Function,* MacMillan, London, 1999.

Kaziro, Y: *Structure and Function of Signal Transducing GTP Binding Proteins,* Blackwell, London, 2001.

Kendrew, J. C. : *The Three-dimensional Structure of a Protein Molecule,* Sinauer, New York, 1991

Kornberg, A. and Baker, T.A. : *DNA Replication,* Freeman, New York, 2002.

Kramer, A. :*The Structure and Function of Proteins Involved in Mammalian Pre-mRNA Splicing,* Gıandada, New York, 2001.

Kurland, C. G. : *Structure and Function of the Bacterial Ribosomes,* Oxford, New York, 2000.

Lake, J. A. : *The Ribosome,* Oxford, New York, 1981.

Lehninger, A. L. : *The Mitochondrion,* Academic Press, New York, 2001

Lewin, Benjamin : *Genes*, Oxford University Press, New York, 2000.

Maxwell, E. S. and Fournier, M.J. : *The Small Nucleolar RNAs*, MacMillan, London, 1995.

McKnight, S. L. : *Molecular Zippers in Gene Regulation*, Princeton, New York, 2002.

Merrick, W. C. : *Overview : Mechanism of Translation Initiation in Eukaryotes*, Longman, London, 2001.

Moore, P. B. : *Molecular Mimicry in Protein Synthesis*, Cambridge, London, 2002.

Noller, H. F. : *Ribosomol RNA and Translation*, Oxford, New York, 1999.

Olson, R. E. : *Enzymatic Synthesis of DNA*, John Wiley, New York, 2001.

Owen, L. :*DNA Synthesis*, Freeeman, San Francisco, 2001.

Pabo, C. O. and R.T. Sauer : *Transcription Factors : Structural Families and Principles of DNA Recognition*, Printice Hall, London, 1993.

Pardee, A. B. : *Molecular Basis of Biological Regulation: Origins of Feedback Inhibition and Allostery*, Arnold, London, 2003.

Pardue, M. L. and J. G. Gall : *Molecular Hybridization of Radioactive DNA to the DNA of Cytological Preparations*, Dekker, New York, 2001.

Paszkowski, J. : *DNA and Histone Methylation in Plants*, MacMillan, London, 2002.

Razin, A. and A. D. Riggs : *DNA Methylation and Gene Function*, Longman, New York, 2002.

Rich, A. and S. H. Kim : *The Three Dimensional Structure of Transfer RNA*, Oxford, New York, 1999.

Robertson, M. : *RNA Processing : the Post-RNA World*, Oxford, New York, 1999.

Rosenberg, M. and D. Court : *Regulatory Sequences Involved in the Promotion and Termination of RNA Transcription*, Cambridge, New York, 2003.

Sager, R. : *Cytoplasmic Genes and Organelles*, Academic Press, New York, 2003.

Schmid, M. B. : *Structure and Function of the Bacterial Chromosome*, Granada, London, 1990.

Spector, D. L. : *The Dynamics of Chromosome Organization and Gene Regulation*, Oxford, New York, 2005.

Stryer, L. : *Biochemistry*, W.H. Freeman & Co., New York, 2002.

Temin, H. M. : *RNA Directed DNA Synthesis*, MacGraw-Hill, New York, 2002.

Turner, B. M. : *Histone Acetylation and Control of Gene Expression*, MacMillan, London, 2002.

Ulrich, B. : *Mechanism and Regulation of Eukaryotic Protein Synthesis*, Oxford, New York, 2002.

Uptain, S. M. : *Basic Mechanisms of Transcript Elongation and its Regulation*, MacMillan, London, 1997.

Vernadsky, V. I. : *Organellar Heredity*, Revan Press, New York, 1978.

Voinnet, O. : *RNA Silencing : Small RNAs as Ubiquitous Regulators of Gene Expression*, MacMillan, London, 2002.

Wallace, D. C. : *Diseases of the Mitochondrial DNA*, Blackwell, New York, 2002.

Wang, T. S. F. : *Eukaryotic DNA Polymerases*, MacMillan, London, 2002.

Watson, J. D. : *Molecular Biology of the Gene*, Oxford, New York, 2002.

———— : *Molecular Structure of Nucleic Acids : a Structure for Deoxyribose Nucleic Acid*, Oxford, New York, 2001.

Wilkins, M. H. F. : *Physical Studies of the Molecular Structure of*

Deoxyribose Nucleic Acid and Nucleoprotein, MacMillan, London, 2002.

Williams, R. C. : *Reconstitution of Active Tobacco Mosaic Virus from its Inactive Protein and Nucleic Acid Components,* Oxford, New York, 1955.

Williamson, B. : *Split Gene Transcription,* Oxford, New York, 2001.

Woese, C. R. : *The Genetic Code : The Molecular Basis for Genetic Expression,* Harper & Row, New York, 2003.

Wuthrich, K. : *Protein Structure Determination in Solution by Nuclear Magnetic Resonance Spectroscopy,* Oxford, New York, 2001.

Yamamoto, T. : *A Bipartite DNA Binding Domain Composed of Direct Repeats in the TATA Box Binding Factor TFHD,* Oxford, New York, 2001.

Yamashita, S. : *Transcription Factor TFHB Sites Important for Interaction with Promoter-bound TFIID,* Springer, London, 2000.

Yams, M. : *Translational Efficiency of Transfer RNAs : Uses of an Extended Anticodon,* MacMillan, London, 2003.

Yanofsky, C. : *The Tryptophan Synthetase System,* Oxford, New York, 2001.

Zawel L., and D. Reinbcrg : *Common Themes in Assembly and Function of Eukaryotic Transcription Complexes,* MacMillan, London, 2002.

Zuckerman, B. : *Initiation of mRNA Translation in Prokaryotes,* Oxford, New York, 1990.

Zurick, D. : *Origin and Evolution of Mitochondrial DNA,* Oxford, New York, 1989.